Alfred Alcock

An Account of the Deep-Sea Brachyura

Collected by the Royal Indian Marine Survey Ship Investigator

Alfred Alcock

An Account of the Deep-Sea Brachyura
Collected by the Royal Indian Marine Survey Ship Investigator

ISBN/EAN: 9783337161163

Printed in Europe, USA, Canada, Australia, Japan

Cover: Foto ©ninafisch / pixelio.de

More available books at **www.hansebooks.com**

AN ACCOUNT

OF THE

DEEP-SEA BRACHYURA

COLLECTED BY THE

ROYAL INDIAN MARINE SURVEY SHIP

INVESTIGATOR

BY

A. ALCOCK, M.B., C.M.Z.S., F.G.S.,

INDIAN MEDICAL SERVICE, SUPERINTENDENT OF THE INDIAN MUSEUM AND PROFESSOR OF ZOOLOGY IN THE MEDICAL COLLEGE, CALCUTTA; FORMERLY NATURALIST TO THE MARINE SURVEY

CALCUTTA:

PRINTED BY ORDER OF THE TRUSTEES OF THE INDIAN MUSEU

1899.

PREFATORY NOTE.

A good many reports, more or less of a preliminary character, have been published in various Journals since 1885, relative to the zoological work of the Marine Survey of India, under the title of *Natural History Notes from H. M. Indian Marine Surveying Steamer 'Investigator,'* and these unofficial reports have been supplemented since 1892 by the official series—published under the authority of the Director of the Indian Marine—of *Illustrations of the Zoology of the Royal Indian Marine Survey Ship 'Investigator.'*

The present volume contains an independent Report upon the Brachyurous Crustacea collected by the 'Investigator,' and it seems advisable to preface it with a short explanation (which has already been published in the *Account of the Investigator Deep Sea Madreporaria*) of the way in which the ship became connected with deep-sea exploration and with the Indian Museum.

In the year 1871 the Council of the Asiatic Society of Bengal appointed Dr. T. Oldham, Dr. F. Stoliczka and Mr. J. Wood-Mason to form a sub-committee to report upon the desirability of moving the Government of India to undertake deep-sea dredging in Indian waters.

The sub-committee drew up an elaborate *Memoir* on the subject, in which definite proposals for deep-sea dredging were embodied: this Memoir was submitted to Government, and a copy of it along with a copy of the letter with which it was forwarded, is published in the *Proceedings of the Asiatic Society of Bengal* for 1871.

The Government received the proposals of the Council of the Asiatic Society with cordial approval: it gave a small grant in aid of carrying them into immediate effect, and when, in 1874, the present Marine Survey Department was established, it sanctioned the appointment, upon the staff of the Survey, of a Surgeon-Naturalist—an appointment that had also been strongly advocated by the organizer and first head of the Department, Commander Dundas Taylor, I. N.

But in the early days of the Survey (1874–1881) neither machinery nor vessels capable of deep-sea research were available, so that Surgeon (now Lieutenant-Colonel) J. Armstrong, I.M.S., the first Surgeon-Naturalist of the Department, had to report that it was "quite impossible to carry into "execution the scheme of deep-sea dredging originally proposed by the Council of the Asiatic Society of "Bengal," and had to confine himself to the Zoology of the shallow-water and littoral, although he did manage to dredge in water as deep as 100 fathoms.

However, in 1876, when it had been decided to construct a special vessel for the accommodation of the Marine Survey, the Council of the Asiatic Society again addressed the Government of India, and asked that provision for deep-sea dredging might not be forgotten in the plans for the new vessel. In reply the Government authorized the Council of the Society to confer with the Dockyard authorities on the subject of such equipment.

The Council thereupon appointed a sub-committee, consisting of Dr. John Anderson, then Superintendent of the Indian Museum, and Messrs. J. Wood-Mason (then Deputy Superintendent of the Indian Museum), W. T. Blanford, H. F. Blanford, and H. B. Medlicott, for the purpose of advising the Dockyard authorities in this direction.

The result of this and other measures was that when, in 1881, the new vessel *Investigator* was ready for sea, she was properly provided with the means of undertaking deep-sea research as opportunity should occur.

Before this, however, Dr. Armstrong had left the Survey, and it was not until the end of the year 1884, when Commander A. Carpenter, R. N., was appointed to the command of the 'Investigator,' and Surgeon (now Major) G. M. J. Giles, I.M.S., to the post of Surgeon-Naturalist, that deep-sea dredging became a recognized, if subordinate, branch of the ship's routine.

Since 1885 the Zoological collections made by the 'Investigator' have been year by year accumulating in the Indian Museum, where, in accordance with the recommendations of the Council of the Asiatic Society of Bengal, they have been deposited.

It must not, however, be supposed that deep-sea dredging occupies a very large part of the attention of the officers of the Survey; since, as a rule, it is only possible when the ship is proceeding to and returning from her systematic surveys of the shores and shallows. It is rarely indeed that as many as twenty deep-sea hauls are made in one year.

From October 1888, when regular records began to be kept, up to the present time, 113 more or less successful hauls have been made in depths of over a hundred fathoms (100–1997 fms.): of these 71 have been under the superintendence of Captain A. R. S. Anderson, I.M.S., who has been Surgeon-Naturalist since 1893.

As regards the 'Investigator' herself, she is a paddle-steamer of 580 tons, and for a few facts as to her history and equipment I may refer to a paper in the *Scientific Memoirs of the Medical Officers of the Army of India* for 1898.

With regard to the contents of the present Report on the *Brachyura*, I may mention that the species that are not here described for the first time have already been noticed in the following papers, relating to the Indian Fauna:—

J. Wood-Mason in Ann. Mag. Nat. Hist., May, 1877, p. 422: in Proc. As. Soc. Bengal, August, 1885: in Journ. As. Soc. Bengal, Vol. LVI, pt. 2, 1887, p. 206, pl. i: and in Ann. Mag. Nat. Hist., March, 1891, pp. 258–270.

A. Alcock in Ann. Mag. Nat. Hist., May, 1894, pp. 400–409: and in *Materials for a Carcinological Fauna of India* in Journ. As. Soc. Bengal, Vols. LXIV, LXV and LXVII, pt. 2, 1895, 1896, 1898.

A. Alcock and A. R. S. Anderson in Journ. As. Soc. Bengal, Vol. LXIII, pt. 2, 1894, pp. 175–185: and in Ann. Mag. Nat. Hist., Jan., 1899, pp. 5–14.

A. R. S. Anderson in Journ. As. Soc. Bengal, Vol. LXV, pt. 2, 1896, pp. 102–106.

Illustrations of the Zoology of the R. I. M. S. "Investigator" by Wood-Mason, Alcock, and Anderson, *Crustacea*, pl. v. xiv. xv. xvi. xvii. xviii. xix. xx. xxi. xxii. xxiii. xxiv. xxv. (xxxvi. xxxvii. xxxix. xl. *in the press*).

The illustrations, like those of the *Account of the Deep Sea Madreporaria*, have been drawn here by Baboos Abhoya Churn Chowdry and Shib Chunder Mondul, and reproduced in collotype by Messrs. Taylor and Francis of London.

A. ALCOCK, Major, I.M.S.,
Superintendent of the Indian Museum.

Corrigenda.

Page 6. *Homola andamanica.* Add Illustrations of the Zoology of the Investigator, Crustacea, pl. xl. fig. 1.

Page 19. For "*Dynomene margarita*" read *Acanthodromia margarita.*

Page 42. After the description of *Physachæus tonsor* insert *Grypachæus hyalinus* (see *Addenda*, p. 83).

Page 50. After the description of *Sphenocarcinus cuneus* insert *Sphenocarcinus auroræ* (see *Addenda*, p. 84).

Page 57. For "Corystoidea" read Cyclometopa, Family *Corystidæ.*

Page 62. After the description of *Orphnoxanthus microps* insert Geryon and *Geryon affinis* (see *Addenda*, pp. 84, 85).

Page 67. After Family *Portunidæ* add Subfamily *Lupinæ.* And for "Goniosoma" read Charybdis (Goniohellenus). And for "*Goniosoma hoplites*" read *Charybdis (Goniohellenus) hoplites.*

Page 68. For "Subfamily *Carcininæ*" read Subfamily *Portuninæ.*

N.B.—1. The species added are from the collections made in the season 1898-99, and were received after the manuscript had been printed. 2. The corrections rest upon an examination of all the *Cyclometopa* and *Dromiacea* in the Indian Museum and are justified in my *Materials for a Carcinological Fauna of India*, No. 4, *The Brachyura Cyclometopa*, and No. 5, *The Brachyura Primigenia* (Journal Asiatic Society Bengal, Part II. 1899). 3. Owing to unavoidable delay in issuing this volume the *Investigator Illustrations* referred to as "in preparation" or "in the press," are now ready and about to be distributed.

An Account of the Deep-sea Brachyura collected by the Royal Indian Marine Survey Ship "Investigator."—*By* A. ALCOCK, M.B., C.M.Z.S., F.G.S., *Superintendent of the Indian Museum and Professor of Zoology in the Medical College of Calcutta: formerly Surgeon-Naturalist to the Indian Marine Survey.*

With two exceptions—namely, a *Homola* that is very possibly identical with the deep-sea *Homola orientalis* of Henderson and a singular species of *Dynomene*—the Brachyura here described came from depths over 100 fathoms.

The species number 53, and belong to 38 genera, and they have been collected by the 'Investigator' during the last fourteen years (1885–1898). Specimens of all, including the types of new species, are in the Indian Museum.

Of these 53 species only the following five are known from other seas:—*Ethusina gracilipes* Miers, *Cyrtomaia Suhmi* Miers, *Platymaia Wyrille-Thomsoni* Miers, *Oxypleurodon Stimpsoni* Miers, *Scyramathia pulchra* Miers. These five are "Challenger" species from the seas of the East Indian Archipelago, but *Ethusina gracilipes* has also, according to Faxon, been taken off the Pacific coast of the Panama region.

To those who expect a faunal list to furnish forth some theory of geographical distribution, our list will appear disappointing.

If, however, we regard genera and not species, the list discloses some suggestive affinities between the Brachyuran fauna of these Seas and of certain parts of the Atlantic area. These affinities may, of course, be taken as merely confirmatory of current views as to the unity of the Deep-sea Fauna; but, seeing that the Brachyura are not generally considered to belong to the true deep-sea (abyssal) fauna, I think it equally probable that they may afford evidence of a former open connexion between the seas in question.

In an *Account of the Investigator Deep-sea Madreporaria* (pp. 2–10) I have discussed this matter at some length, so that here I need speak only of the supplementary evidence that the deep-sea Brachyura appear to furnish of this connexion.

Of the 38 genera that are at present known to compose the Brachyuran fauna of the Indian depths the following 21 are also known to occur in other seas:—*Homola*, *Dynomene*, *Calappa*, *Mursia*, *Randallia*, *Ethusa*, *Ethusina*, *Lyreidus*, *Echinoplax*, *Cyrtomaia*, *Platymaia*, *Sphenocarcinus*, *Oxypleurodon*, *Scyramathia*, *Maia*, *Trichopeltarium*, *Trachycarcinus*, *Goniosoma*, *Pilumnoplax*, *Carcinoplax*, *Pinnoteres*.

Homola. This genus, which is represented in India by three good species, was long regarded as characteristic of the Mediterranean. The Mediterranean species (*H. barbata*) is now known to occur in the West Indies and neighbouring coasts of N. America, and a form that is probably only a variety of it—described in the sequel as *H. andamanica*—has lately been taken in the Martaban end of the Andaman Sea at 79—90 fms. This *H. andamanica* may possibly be the same as Henderson's *H. orientalis* from the Banda and Sulu Seas.

Furthermore, of the 3 Indian species of *Homola*, one—described in the sequel as *H. profundorum*—is closely related to the Mediterranean *H. Cuvieri* in Wood-Mason's genus (preferably subgenus) *Paromola*.

Dynomene ranges across the whole Indo-Pacific from Mauritius to California.

Calappa is a shallow-water genus: the species are distributed all over the Indo-Pacific, from Africa to California, and are also represented in the West Indies, the Mediterranean, and on the west coast of Africa as far as the Cape.

Mursia. The geographical distribution is not remarkably different from that of *Calappa*, except that it does not occur in the Mediterranean.

Randallia seems to be confined to the Indo-Pacific.

Ethusa, of which Ethusina is a subgenus, extends from the West Indies and neighbouring coasts of N. America to the Azores and the Mediterranean. In the Indo-Pacific it extends from the Arabian Sea to Japan and the Philippines, to Fiji, and to California and the Pacific coast of Panama.

Lyreidus. One species is common in Indian waters at 200–400 fathoms: another belongs to the North American Atlantic slopes at 100 fathoms: a third species belongs to the Japanese fauna, and also extends into Australasian waters.

Echinoplax, Cyrtomaia and Platymaia appear to be confined to the Indo-Pacific. But I do not think that *Cyrtomaia* is really different from *Echinoplax*, or that either should be separated from the Atlantic and Mediterranean *Ergasticus* of A. Milne-Edwards.

Sphenocarcinus, one species of which occurs in the Andaman Sea at 161–250 fathoms, is a genus originally discovered in the Caribbean Sea at 100 fathoms. A third species, however, has been described by Miss Rathbun from shallower water in the Gulf of California. *Oxypleurodon* hardly differs from *Sphenocarcinus*.

Scyramathia is a North-Atlantic genus (European, N. American and Caribbean) that is well represented in Indian Seas. Other species are known from the Philippines and from the Galapagos.

Maia. In the time of H. Milne Edwards this was supposed to be a genus peculiar to the seas of Europe (Mediterranean, British Seas, North Sea). One species is common to Japan and the East Indies, one has been described from Australasia, and recently two more have been added from the Oriental region.

Of the 2 Corystoid genera that occur in Indian deep waters:

TRICHOPELTARIUM is only known elsewhere from the Caribbean Sea, while the closely allied TRACHYCARCINUS is from the Pacific slopes of Central America.

GONIOSOMA, PILUMNOPLAX and CARCINOPLAX may be regarded as shallow-water Indo-Pacific genera, *Pilumnoplax* occurring also in the South Atlantic.

Of the genera here referred to as peculiar (so far as is known) to Indian Seas, two require special notice in connexion with this question of geographical distribution.

The first of these is the Dromioid genus ARACHNODROMIA. This is so extremely like Prof. A. Milne Edwards' *Homolodromia* from the Caribbean Sea, that I should have taken it almost for the same species had not M. Milne Edwards stated that the form described by him has neither antennulary nor special orbital fossæ. Our species has common antennulo-orbital fossæ not essentially different from those of *Dromia*.

The second is BENTHOCHASCON, which is very intimately related to the Gulf of Mexican *Bathynectes*. The last named genus, however, is also known from the Eastern Pacific.

The Composition of the Brachyuran Fauna of the Indian Oligo-benthos.

This, so far as is known, is as follows:—

Catometopa	7	species, or	13·2 per cent.
Cyclometopa (*including* Corystoidea)	8	,, ,,	15·0 ,, ,,
Dromidea (Dromidæ and Homolidæ)	9	,, ,,	16·9 ,, ,,
Oxystoma (*including* Raninidæ)	14	,, ,,	26·4 ,, ,,
Oxyrhyncha	15	,, ,,	28·3 ,, ,,

The large proportion of Oxyrhynchs and the small proportion of Cyclometopes and of Catometopes (due regard being paid to the relative proportions of these groups in their entirety) is interesting to those who do not accept the view that the Oxyrhynchs as a whole are the highest Crustacean developments.

The very large proportion, relative to the size of the group, of *Dromidea* is also of much interest.

The Bathymetric Range of the Indian Deep-sea Brachyura.

Only one species—*Ethusa investigatoris*—has been dredged in 1200–1300 fathoms.

The following three species occur between 800 and 1,000 fathoms:—

Hypsophrys superciliosa	740–931	fms.
Ethusa desciscens	931	,,
Ethusa gracilipes	836	,,

The three following species occur between 500 and 800 fathoms :—

Paromolopsis Boasi	597 fms.
Ethusa indica	719 „
Scyramathia pulchra	561 „

The following 18 species occur between 400 and 500 fathoms.

Dro-mid.	*Homola profundorum*	430	fms.
	Hypsophrys longipes	430	„
	Arachnodromia Baffini	430	„
Oxyst.	*Randallia pustulosa*	406	„
	Cymonomops glaucomma	405	„
	Lyreidus Channeri	406	„
Oxyrh.	*Physachaeus ctenurus*	406	„
	Cyrtomaia Suhmi	430	„
	Platymaia Wyville-Thomsoni	405	„
	Encephaloides Rivers-Andersoni	406	„
	Scyramathia Rivers-Andersoni	406–430	„
Cancer.	*Trachycarcinus glaucus*	430	„
	Benthochascon Hemingi	405	„
Catom.	*Carcinoplax longipes*	430	„
	Pilumnoplax Sinclairi	430	„
	Psopheticus stridulans	419	„
	Hephthopelta lugubris	490	„
	Pinnoteres abyssicola	430	„

Thus rather more than half the species of Indian deep-sea crabs have been taken only in depths between 400 and 100 fathoms. Moreover, a good many of the species that range into the greater depths mentioned above also occur in shallower water.

On the other hand, with the single doubtful exception of *Doclea ovis*, we have no instance of a true shallow-water crab being taken so far out as the hundred-fathom line.

The majority of the species here described have already been noticed either in the *Annals and Magazine of Natural History* or in the *Journal of the Asiatic Society of Bengal*, and a good many of them have been figured in the *Illustrations of the Zoology of the Investigator*.

My predecessor, Professor J. Wood-Mason, was the author of the first published notices and figures,—in the *Journal of the Asiatic Society of Bengal*, for 1887, pt. 2, pp. 206–209, pl. i, in the *Annals and Magazine of Natural History* for March, 1891, pp. 258–270, fig. 5, and in the *Illustrations of the Zoology of the Investigator*, 1892, Crustacea, pl. v, in which contributions fourteen valid new species and seven new genera are included.

List of the Investigator Deep-sea Brachyura.

Homolidæ	*Homola andamanica,*	79—90 fathoms.
	Homola megalops,	142–400 fathoms.
	Homola profundorum,	430 fathoms.
	Paromolopsis boasi,	360–597 fathoms.
	Hypsophrys superciliosa,	740–931 fathoms.
	Hypsophrys longipes,	430 fathoms.
Dromidæ	*Arachnodromia baffini,*	238–430 fathoms.
	Sphærodromia kendalli,	112 fathoms.
	Dynomene margarita,	75 fathoms.
Oxystoma.	*Calappa exanthematosa,*	91–112 fathoms.
	Mursia bicristimana,	142–400 fathoms.
	Mursia aspera,	210 fathoms.
	Randallia lamellidentata,	350 fathoms.
	Randallia pustulosa,	220–406 fathoms.
	Parilia alcocki,	70–250 fathoms.
	Pariphiculus coronatus,	112 fathoms.
	Ethusa indica,	240–719 fathoms.
	Ethusa pygmæa,	168–240 fathoms.
	Ethusa andamanica,	168–290 fathoms.
	Ethusina gracilipes,	836 fathoms.
	Ethusina investigatoris,	1,200–1,300 fathoms.
	Ethusina desciscens,	265–931 fathoms.
	Cymonomops glaucomma,	265–405 fathoms.
	Lyreidus channeri	200–406 fathoms.
Oxyrhyncha	*Physachæus ctenurus,*	185–406 fathoms.
	Physachæus tonsor,	271 fathoms.
	Echinoplax pungens,	112–250 fathoms.
	Echinoplax rubida,	90–177 fathoms.
	Cyrtomaia suhmi,	430 fathoms.
	Platymaia Wyville-thomsoni,	130–405 fathoms.
	Encephaloides armstrongi,	60–100 fathoms.
	Encephaloides rivers-andersoni,	406 fathoms.
	Sphenocarcinus cuneus,	161–250 fathoms.
	Oxypleurodon stimpsoni,	180–217 fathoms.
	Scyramathia pulchra,	130–561 fathoms.
	Scyramathia rivers-andersoni,	406–430 fathoms.
	Scyramathia beauchampi,	193–210 fathoms.
	Scyramathia globulifera,	130–240 fathoms.
	Maia gibba,	250 fathoms.
Corystidæ.	*Trichopeltarium ovale,*	180–217 fathoms.
	Trachycarcinus glaucus,	430 fathoms.
Cyclometopa.	*Orphnoxanthus microps,*	105–350 fathoms.
	Platypilumnus gracilipes,	188–220 fathoms.
	Nectopanope rhodobaphes,	98–102 fathoms.
	Sphenomerides trapezioides,	130–290 fathoms.
	Benthochascon hemingi,	185–405 fathoms.
	Goniosoma hoplites,	80–110 fathoms.

Catometopa.		
	Pilumnoplax sinclairi,	430 fathoms.
	Carcinoplax longipes,	220–430 fathoms.
	Psopheticus stridulans,	173–419 fathoms.
	Camatopsis rubida.	194 fathoms.
	Hephthopelta lugubris,	490 fathoms.
	Ptenoplax notopus,	100–250 fathoms.
	Pinnoteres abyssicola,	430 fathoms.

Brachyura anomala.

Family *Homolidæ.*

Homola, Leach.

Homola, Leach, Trans. Linn. Soc., Vol. XI. 1815, p. 324, and Zool. Miscell. Vol. II. p. 82, pl. lxxxviii: Latreille, [Nouv. Dict. d'Hist. Nat.], and in Cuvier's Regne Animal, ed. 1829, p. 67: Desmarest, Consid. Gen. Crust. p. 133: Risso, Hist. Nat. Europ. Merid. Vol. V. pp. 34-35: Roux, Crust. de la Mediterranée text of pl. vii. Milne Edwards, Hist. Nat. Crust. II. 181: Dellaan, Faun. Japon. Crust. p. 105: Dana U. S. Expl. Exp Crust. pt. I. p. 403: Heller, Crust. Sudl. Europ. p. 148: Henderson, "Challenger" Anomura, p. 18: Ortmann, Zool. Jahrb. Syst. etc. VI. 1892, pp. 540 and 542: A. Milne Edwards and Bouvier, "Hirondelle" Brachyures et Anomures (Monaco 1894) p. 60.

Carapace deep, longer than broad, quadrilateral or urn-shaped, with deep vertical sides, the gastric region well demarcated and occupying the anterior half of the carapace, the linea anomurica distinct and dorsal.

Front narrow, forming a rostrum, which is either entire or bifid at tip and has a spine, often of large size, on either side of its base.

The orbits are quite incomplete and do not even conceal the eye-stalks, and the eyes, which project far outside them, are retractile against the sides of the carapace. The eye-stalks are long and are composed of two joints, a slender basal joint, and a swollen terminal joint that carries the eye.

There are no antennulary fossæ: the antennules are large and their basal joint is subglobular.

The antenna-peduncle is long slender and freely movable from its base, and is inserted in almost the same plane as the antennules: it is composed of 4 joints, the basal one of which has a strong "auditory tubercle." The flagellum is extremely long.

The epistome is fairly or very distinctly marked off from the palate. The buccal cavern is subquadrilateral, but broader in front than behind; the expiratory canals are very well defined. The external maxillipeds are subpediform.

The chelipeds are rather slender and generally somewhat spiny. The legs are long and more or less compressed and spiny, the last pair are dorsal in position, and are subcheliform, but have the propodite dilated near the basal end and never twice the length of the dactylus.

The abdomen of the male consists of seven separate segments and is rather broad.

Wood-Mason (Ann. Mag. Nat. Hist. March 1891, p. 267) separated *Homola cuvieri* from *Homola barbata*, basing his opinion on the form of the carapace, the position of the *linea anomurica*, and the form of the terminal joints of the last pair of legs.

He appears to have regarded the hollowed out portion of the carapace against which the eye can rest in retraction as a commencing orbit,—a view that seems to me to be more than doubtful.

I agree with him, however, that the differences between the two forms are of more than specific importance, and I am inclined to maintain *Paromola* as a subgenus. The form described by me in Ann. Mag. Nat. Hist. May 1894, p. 408, as *Homola megalops* now seems to me to be also worthy of subgeneric rank.

These three sections of the genus *Homola* may be thus characterized :—

1. Homola. Carapace square-cut, its broadest part being in front, across the middle of the gastric region: the *linea anomurica* rather inconspicuous, keeping close to the lateral border. Rostrum a non-cylindrical bifid tooth, with a smaller spine on either side of its base. 2nd joint of antenna-peduncle having its antero-external angle produced to form a spine. Palate distinctly delimited from the epistome everywhere except in the middle line. The last pair of legs reach to the end of the carpus of the preceding pair.

Type *Homola barbata* (Herbst.)

Homolax. Carapace urn-shaped, its greatest breadth being behind, across the middle of the branchial regions: the *linea anomurica* conspicuous, running well inside the lateral border. Rostrum as in *Homola*. 2nd joint of antenna peduncle having its antero-external angle acute, but not spiniform. Palate as well demarcated from the epistome in the middle line as it is elsewhere. The last pair of legs reach beyond the end of the carpus of the preceding pair.

Type *Homola megalops*, Alcock.

Paromola Wood-Mason. "Carapace decidedly macrurous in form," its greatest breadth being behind: the *linea anomurica* very conspicuous and well inside the lateral border. Rostrum a simple cylindrical spine of large size flanked on either side by a single spine of equal or greater size. 2nd joint of antenna-peduncle not produced or specially acute at the antero-external angle. Palate everywhere well demarcated from the epistome. The last pair of legs not reaching beyond the end of the merus of the preceding pair.

Type *Homola cuvieri*, Roux.

Subgenus *Homola*.

Homola andamanica, n. sp.?

This may, very possibly, prove the same as *Homola orientalis* Henderson, though it cannot be quite reconciled with the description, still less with the figure, of that species.

In any case it is probably only a variety of *Homola barbata* with 3 good specimens of which—representing both sexes—it has been compared. The only differences between it and *H. barbata* are the following:—

The eyes are more reniform. The second spine of the lateral border is just behind the hepatic region. There are spines on the posterior border of the meropodites of all four pairs of walking legs.

Carapace elongate-subquadrilateral, its greatest breadth is across the middle of the gastric region, behind which point its sides are quite straight and vertical: it is well calcified, and, like all other parts except the antennary flagella, is covered with short soft but stiff hairs that are not thick set enough to form a coat of concealment.

Rostrum a depressed grooved tooth, bifid at tip. Four spines on the anterior border of the carapace, namely, one on either side of the rostrum, one at either supra-orbital angle.

Lateral borders of dorsum of carapace straight, very slightly convergent, spinate; the first spine, which stands alone on the hepatic region, is of pre-eminent size, the second though much smaller than the first is much larger than any of the others.

Gastric region very well demarcated, armed with nine large spines—three in a triangle on either median area, one on either lateral area, and one on the hinder part of the central area.

Some spines on the subocular, subhepatic, and pterygostomian regions—largest on the subocular region, where they are definitely arranged in two crescentic rows. Two spines, one *beside* the other, on the carapace outside the antenna-peduncle, in addition to the spinuliform suborbital angle.

Eyes somewhat reniform.

Chelipeds slender, but distinctly stouter than the legs, more hairy than the carapace, especially along the edges of the joints. Upper and lower borders of arm spiny; wrist with rows of spines on the outer surface and a spine or two at the inner angle; lower border of hand spiny, upper border of hand denticulate, cutting edges of fingers sharp, entire.

Legs compressed, their edges plumed with short bristles, with long bristles interspersed. The second and third pair, which are a dactyl-length longer than the first, are not quite $2\frac{1}{3}$ times the length of the carapace: in all three pairs both edges of the merus are armed with stout spines—at least in the distal half, and the posterior border of the propus and dactylus with compressed articulated spines which are distant and acicular on the propus but stout very regular and close-set on the dactylus.

The subcheliform fourth pair of legs reach very slightly beyond the end of the carpus of the preceding pair: the merus has 3 or 4 spines on the lower border and a terminal spine on the upper border, the claw-like dactylus closes against a bunch of spines on the near end of the propus.

From the Andaman Sea, 79–90 fms.

Subgenus *Homolax.*

Homola megalops, Alcock.

Homola megalops, Alcock, Ann. Mag. Nat. Hist., May 1894, p. 408: Illustrations of the Zoology of the R. I. M. S. 'Investigator,' Crustacea pl. xiv. figs. 1, 1a.

Carapace urn-shaped, its greatest breadth is across the middle of the branchial region; its sides, and still more the spinulate lateral borders of its dorsum, are elegantly curved; the hairs that cover it are so inconspicuous as to be recognizable only with a lens.

Rostrum a depressed grooved tooth, entire, or emarginate at tip. Four spines on the anterior border of the carapace arranged as in *H. barbata.*

The only enlarged spine of the lateral border stands alone on the hepatic region.

Nine spines on the gastric region—two immediately behind the spines at the base of the rostrum, the other seven in an open S-shaped curve across the middle of the region.

A single row of spines on the subocular region, which region is remarkably hollowed for the reception of the retracted eye. Two spines, one above the other, on the carapace beside the antenna-peduncle, in addition to the bluntly-dentiform suborbital angle.

Eyes reniform, very large, their major diameter being one-sixth the breadth of the carapace.

Chelipeds slender, their arms and wrists distinctly slenderer than the meropodites of the legs: in the adult male they do not reach half-way along the merus of the first pair of legs: they are covered with a short inconspicuous velvet, with hardly any long bristles on the edges of the joints: they are armed much as in *H. barbata*, but the *upper* border of the hand is spiny and the lower border faintly denticulate. The fingers, which have a sharp entire cutting-edge, are as long as the rest of the hand.

The legs have the surface—especially the dorsal surface—of most of the joints covered with a close short velvet, but have few or no bristles along their edges. The 2nd and 3rd pair, which are nearly a dactylus longer than the first, are nearly three times as long as the carapace: the subcheliform 4th pair reach beyond the end of the carpus of the preceding pair. The first three pair have the anterior edge of their greatly compressed meropodite closely spinate, and the posterior edges of that joint and the ischium closely spinulate; their last three joints have the edges smooth, except for a few small jointed spinules at the base of the posterior border of the dactylus. The last pair of legs have the posterior edge of their subcylindrical meropodite closely spinate and have only a single terminal spine on the upper edge, the carpus has a strong terminal spine on its posterior border, and the propus has a salient group of spines behind the middle of its posterior border forming a subcheliform stump for the serrated posterior edge of the claw-like dactylus.

Colour in life salmon-pink.

Andaman Sea, 188–220 fms., a male and a female; 370–419 fms., 3 males and 3 females. Bay of Bengal, off Coromandel Coast, 145–250 fms., a male and a female. Gulf of Manár, off Colombo, 142–400 fms., 2 young males.

Dimensions of carapace of a full-grown specimen 41 millim. long, 36 millim. broad.

The gills are fourteen in number on either side, exclusive of a quite rudimentary posterior arthrobranch to the penultimate pair of legs.

Subgenus *Paromola.*

Homola profundorum, Alcock and Anderson, Plate I, fig. 2.

Carapace very decidedly macruriform, deep, ovoid-triangular, broadest abaft the middle of the branchial region, tapering to an acutely-spiniform rostrum of which the length is about a third that of the rest of the carapace. Diverging from either side of the base of the rostrum is a spine of similar form and size. The only other elevations on the carapace are a hepatic spine just behind the hollow for the retracted eye, an antennal spine just outside the antennal base, and a blunt denticle near the middle of the ill-defined lateral border.

The gastric region is well delimited, and the *linea anomurica* is broad conspicuous and dorsal.

The stout cylindrical terminal joint of the eye-stalks is longer than the slender basal joint, the eyes are of good size, well pigmented, and hemispherical.

The chelipeds are slender but are stouter than the legs; the arm has the outer lower border spinate and, on the upper border, a few spinules and a strong terminal spine; both the inner and the outer angles of the wrist are armed with a strong spine, the fingers are much shorter than the hand and have the cutting-edge entire.

The legs are slender and subcylindrical, the 2nd and 3rd pair, which are slightly longer than the first, are at least three times the length of the carapace. In the first 3 pair there are a few distant spines and a strong terminal spine on the anterior border of the merus, a few articulating spinules at the far end of the posterior border of the propodite, and a comb of articulating spines along the posterior border of the dactylus—the last joint being but half the length of the last but one. The dorsal fourth pair of legs are far slenderer than the others and do not reach the end of the merus of the preceding pair: their propodite is triangular, owing to the expansion of its posterior border, and opposes a sharply-serrated edge to the less strongly toothed posterior border of the short dactylus—the parts being cheliform rather than subcheliform.

The body and appendages are coated with very short distant bristles which do not conceal the surface: there are some longer and thicker bristles along the edges of the chelipeds, and a very few scattered hairs along the edges of the legs.

Three young females from off the Travancore coast, 430 fms.

The carapace of these is about 13 millim. long, and about 9 millim. in greatest breadth.

PAROMOLOPSIS, Wood-Mason.

Resembles *Homola* but differs in the following important particulars :—

The carapace is "more brachyurous:" it is urn-shaped and *depressed*, its sides being far from vertical and being overhung by the sharply defined lateral borders. The hepatic region is elongate and advanced, so that the hepatic spine is on a level with the spines of the anterior border, and helps to form a very decided "orbit." The buccal cavern is scarcely broader in front than behind.

In other respects it agrees with *Homola* and more particularly with the subgenus *Homolax*.

The branchial formula is the same as that of *Homola*.

Paromolopsis boasi, Wood-Mason.

Paromolopsis boasi, Wood-Mason, Ann. Mag. Nat. Hist. March 1891, p. 268 and fig. 5.

Every exposed surface of the body and appendages, excepting only the flagella of the antennæ, is covered with an even, velvet-like, tomentum.

Carapace ending in a short triangular rostrum with an upturned tip, its greatest breadth, which is across the middle of the branchial regions, is equal to its length without the rostrum. Unlike the species of *Homola*, the lateral border is well-defined throughout, is carinated, is co-extensive with the length of the carapace, and ends in a large triangular hepatic spine the tip of which is on a level with the tips of the spines of the anterior border: these are four in number, one on either side of the rostrum and one at either outer orbital angle.

There is an antennal spine and spinule, there are some definitely-placed nodular swellings on the well defined gastric region, and the surface of the denuded carapace is granular, but there are no spines other than those mentioned.

The swollen terminal joint of the eyestalk is rather longer than the slender basal joint: eyes of good size, well pigmented, hemispherical, retractile into a very decided hollow in the front wall of the hepatic region.

The 2nd joint of the antenna-peduncle is not produced or acute at the antero-external angle; the antennary flagellum is much longer than the carapace.

Chelipeds (in the adult female and young male) short, just reaching beyond the end of the carpus of the first pair of legs: the arm is slenderer than the corresponding joint of the first three pair of legs: the fingers are longer than the hands: none of the joints are spinate.

The second and third legs, which are longer than the first by their dactylus, and longer than the fourth by their merus and dactylus, are 3 times the length of the carapace. In the first three pair of legs the anterior border of the meropodite is armed with large spines, but the other joints are unarmed: the dactylus is slender, curved, and of great length, being hardly shorter than the preceding joint.

In the subcheliform, dorsal, fourth pair the anterior border of the merus ends in a spine and the posterior border of the merus is spiny throughout, the propus is much dilated and toothed at its basal angle posteriorly, so as to be *l*-shaped and has one or two spines on the undilated portion of its posterior border, and the dactylus is short and is toothed along the posterior border.

The abdomen of the male consists of seven segments.

The carapace of an adult female is 45 millim. long and 43·5 millim. broad.

The colours in life vary from red to bluish-pink.

In the Indian Museum are a large female and three young females from off the Andamans, 480–500 fms., 498 fms. and 561 fms.; a young male a large adult female and four young females from off the Travancore coast, 406 and 430 fms.; a large female with eggs from off the Laccadives, 360 fms.; and a young female from off Colombo, 597 fms.

Hypsophrys, Wood-Mason.

Carapace deep, longer than broad, quadrilateral or ovate-oblong, with deep vertical parallel sides, the gastric region well delimited and occupying its anterior half, the *linea anomurica* dorsal, distinct or indistinct.

Front narrow, forming a simple or bifid rostrum which has a spine on either side of its base.

The orbits do not afford any concealment to the eyes, but form, on either side of the rostrum, a broad concave facet sharply marked off from the rest of the carapace by a ridge that arches round dorsally from the rostrum to the antennal spine: at the upper and inner angle of this facet is a well defined hollow that catches the knee of the 2nd and 3rd joints of the antennulary peduncle when flexed. The eyes are well formed: the terminal joint of the eyestalk is barrel-shaped much as in *Homola*, but the slender basal joint is short or obsolescent, so that the eyes do not appreciably project beyond the edge of the orbital facet.

The antennules and antennæ are identical with those of *Homola*.

The mouth-parts also are very like those of *Homola*, but as the outer border of the merus of the external maxillipeds is hardly at all expanded these appendages are even more pediform than in *Homola*.

Chelipeds slender, spiny, equal. Legs of the first three pair long, with broad compressed meropodites. Fourth pair of legs, dorsal in position, short, very slender, cheliform, their dactylus, which is many times shorter than their propodus, shutting down against and co-terminous with the slightly expanded distal end of the propodus.

The abdomen of the male consists of seven separate segments.

In general form *Hypsophrys* resembles *Homola barbata*, but it differs from *Homola* in the following particulars:—

1. The eyestalks are like those of *Dromia*, the long slender basal joint of *Homola* being reduced to next to nothing.

2. Though there are no true orbits there are distinct orbital facets, and the homologies of these with the orbits of *Dromia*—in respect both of conformation and of common use for eyes and antennules—are unmistakeable.

3. The external maxillipeds are unequivocally pediform, the merus being hardly broader than the ischium.

4. The fourth (last) pair of legs have the subchelæ or chelæ quite different in form: the propodite is long and is slightly expanded at its distal end, and the dactylus is a minute joint, ever so much smaller than the propodite, that shuts down against the *distal border* of the latter like the blade of a knife.

Wood-Mason, who regarded the plane or hollow surface on the antero-lateral wall of the carapace of *Homola*, against which the eye can be retracted, as a commencing orbit, said that *Hypsophrys* has no orbits; and this is quite correct if the surface referred to be really an orbit.

But if we compare the carapace of *Hypsophrys* with that of *Dromia*, and regard the orbit as the hollow included between the rostrum and the antennal spine, then *Hysophrys* has far better orbits than *Homola*, for the space in question is a distinct depression sharply marked off from the rest of the carapace by a ridge.

Hypsophrys superciliosa, Wood-Mason.

Hypsophrys superciliosa, Wood-Mason, Ann. Mag. Nat. Hist. March 1891, p. 269: Illustrations of the Zoology of the "Investigator," Crust. pl. xiv. figs. 4, 4a, 1895.

Rostrum simply pointed. Linea anomurica rather indistinct.

Four small spines or teeth on the anterior (orbital) border of the carapace, two being far apart at the base of the rostrum and one at either outer orbital angle. Two, or all four, of these teeth may be obsolescent or obsolete.

Lateral borders of dorsum of carapace not defined, except by a single isolated spine on the hepatic region. Gastric region sharply subdivided into three subregions, of which the lateral are somewhat nodular. Two or three spines on the subhepatic and suborbital region, the innermost of which is "antennal," also sometimes a few spinules.

Eyes well formed and facetted, but pale. Antennal flagella about half again as long as the carapace.

The pediform external maxillipeds have their surfaces and edges devoid of spines.

Chelipeds slender, but much more massive than the legs, about half a hand-length shorter than the first pair of legs in the adult male: spines and spinules in rows on edges and on both inner and outer surfaces of arms, wrists and hands: fingers about three-fourths the length of the palm.

The second pair of legs, which are slightly longer than the first and third and considerably more than twice the length of the fourth, are slightly more than three times the length of the carapace.

In the first three pair the meropodites are compressed with the anterior border spiny and the posterior border much less strongly and profusely spiny, the other joints are slender and unarmed except for a few articulating spinelets at the far end of the posterior border of the propodite and in the basal half of the posterior border of the dactylus, the dactylus is slightly shorter than the propodite.

The fourth (dorsal) pair are very slender and are unarmed except at their cheliform ending: their propodite is many times longer than the dactylus.

The terminal joint of the male abdomen is bluntly triangular.

There are some soft bristles on the chelipeds, and a few on the legs, and some very short and inconspicuous hairs on the carapace.

Colours in life, pink.

The carapace of a large egg-laden female is 19 millim. long and 15 millim. broad.

This species has frequently been taken in the Laccadive Sea and in the sea to the north of the Laccadives at depths ranging from 740 to 931 fms., on soft bottoms.

In the Indian Museum are more than 30 specimens representing both sexes adult and in young stages.

Hypsophrys longipes, Alcock and Anderson. Plate I, fig. 1.

Rostrum deeply bifid. Linea anomurica distinct.

Four large spines on the anterior border of the carapace—two close together at the base of the rostrum, one at either orbital angle.

Lateral borders of dorsum of carapace well defined, spinulate; the ridge on the side-wall of the carapace that defines the branchial regions anteriorly is also spinulate. A row of spines on the hepatic region, the largest of which is on the lateral border of the carapace and has a spine dorsad of it.

Gastric region obscurely subdivided, each lateral subregion is armed with 5 or 6 large spines, while on the median region there is a central spine sometimes followed by a row of spinules. Subhepatic and suborbital region with numerous large spines, one of which is "antennal."

Eyes well pigmented. Antennal flagella more than twice the length of the carapace.

Rows of spinules on the exposed surface of the ischium merus and exognath of the external maxillipeds, and a row on the basal joint of the antennules.

Chelipeds slender, reaching not far beyond the end of the carpus of the first pair of legs, the arm and wrist not stouter than the meropodites of the first 3 pair of legs; spinate and spinulate as in the preceding species; fingers as long as the hand.

The second and third pair of legs, which are slightly longer than the first and three times as long as the fourth, are four times the length of the carapace. In the first three pair of legs the merus is compressed and has its anterior border spinate and its posterior borders spinulate, the posterior border of the propodite carries a few distant articulating spinelets, and the dactylus—which is about two-thirds the length of the preceding joint—has a close comb of articulating spines along its posterior border.

The fourth (dorsal) pair, which are extremely slender, have the posterior border of the merus strongly spinate: the propodite is several times longer than the minute dactylus.

The terminal joint of the male abdomen ends acutely.

Hairs and bristles are sparsely present just as in the preceding species.

The carapace of a large egg-laden female is 38 millim. long and 30 millim. broad.

Eleven specimens, representing adults and young of both sexes, were lately dredged off the coast of Travancore at 430 fms., on a bottom which, though muddy, was covered with coral of the genera *Caryophyllia*, *Desmophyllum*, *Solenosmilia*, and *Lophohelia*. According to Dr. A. R. Anderson, the present Naturalist to the Survey, nearly half a ton of coral was brought up at this haul.

Family *Dromidæ.*

DROMIA, Fabr., Edw.

Sub-genus SPHÆRODROMIA, *nov.*

Carapace globose and pilose: palate with a low and inconspicuous ridge on either side: rostrum subacutely bilobed: the sternal grooves of the female are very short, ending well behind the level of the genital openings: chelipeds and legs as in *Dromia rumphii*.

Dromia (Sphærodromia) Kendalli, (Alcock & Anderson).

Dromidia Kendalli, Alcock & Anderson, J. A. S. B. Vol. LXIII. pt. 2, 1894, p. 175: Ill. Zool. Investigator, Crust, pl. xxiv. figs. 1, 1a.

Carapace globose, its length and breadth equal, closely covered—like the appendages and the other exposed surfaces of the body—with dense short yellowish fur, many of the hairs being club-shaped; its surface smooth, except for a few vesiculous granules on the pterygostomian regions and on the posterior part of the side-wall. Cardiac region clearly defined; so also is the cervical groove, especially in the part of its course that traverses the side-walls of the carapace.

The front consists of two broadly triangular teeth and, on each side, breaks into the roof of the orbit so as to imperfectly divide that cavity into two fossæ—one corresponding with the eyestalk the other with the eye. There is no median tooth above the junction with the epistome.

The lateral borders of the carapace are arched, and are entire except for a few granules visible only with a lens; the anterior half of each border is carinate. Posterior border concave.

The common antennular-orbital cavities are deep, affording complete concealment to the retracted parts: the tooth at the outer lower orbital angle is broadly and bluntly triangular and forms more than a third of the floor of each cavity. Eyes well formed, but decidedly deficient in pigment. Antennal flagella long.

The external maxillipeds, though broadly operculiform, do not meet against the raised epistomial margin, but leave there a wide gap.

Chelipeds, in the female, about $1\frac{1}{3}$ times the length of the carapace; when denuded, there are distant vesiculous granules on the upper and lower edges of the arms, on the upper and outer surfaces of the wrists, and on the hands everywhere except the lower part of the inner surface: fingers hollowed, the immobile finger more strongly toothed than the dactylus.

The first two pairs of legs are stout and about as long as the chelipeds. The last two pairs are slender, and are about half the length of the other legs, and each ends in a small claw-like dactylus which is opposed to two or three little claw-like spines at the end of the propodite.

The sternal grooves of the female are very short, ending *well behind the level of the genital openings*. The female has 5 pair of abdominal legs of the usual form.

A single female with a carapace 18 millim. in diameter from the Bay of Bengal, off Nellore coast, 112 fms.

ARACHNODROMIA.

Carapace elongate-oblong but somewhat broader behind than in front, deep, inflated, tomentose, its texture thin but well calcified: two creases break either lateral border, the posterior one being the more distinct and being continued to the cardiac region (= cervical groove).

The front is horizontal, prominent, and bifid.

The antennule and eye of either side are completely retractile into a common deep fossa (just as in *Dromia*) which affords them complete protection. As in *Dromia*, the floor of this common antennular-orbital fossa is formed by a sub-ocular ("antennal") tooth in contact with the basal joint of the antenna, and, as in *Dromia*, the outer wall of the orbit is breached by a wide gap. The orbital portion of the fossa, which is loosely filled by the eyes, has the hollow for the eyes much deeper than the hollow for the eyestalk. The eyestalks are long and slender, the eyes small but perfectly formed and well pigmented.

The two basal joints of the antennæ, which are quite freely movable, largely fill the gap in the lower wall of the orbit, and lie in the same plane with the antennules; the second joint has its antero-external angle produced to form a coarsish spine: the antennal flagella are longer than the carapace.

The palate is particularly well demarcated from the epistome and is rather broader in front than behind: the ridges that define the expiratory canals are very distinct. The epistome is in the closest possible contact with the front, but without complete fusion. The external maxillipeds are distinctly operculiform, but owing to the moderate expansion of the merus and to the coarseness of the palp, they have a slight pediform cast: they close the buccal cavern, but not so tightly as in *Dromia*.

The chelipeds are equal and are rather slender, though considerably stouter than the legs: the fingers are well calcified and are hollowed *en cuillère*, the tip of the dactylus shuts into a notch in the tip of the opposed finger.

The legs are cylindrical: the first two pairs are very long, the last two are short, subdorsal in position, and cheliform rather than subcheliform.

The sternal grooves of the female end opposite the openings of the oviducts, without tubercles.

The abdomen of both sexes consists of seven distinct segments. In both sexes the pleuræ of the 3rd–6th abdominal somites are remarkably free and independent (*i. e.* not in contact with those in front and behind) and the last abdominal tergum is nearly as long as the preceding five combined. In the male this last tergum is marked in a most suggestive way (see figure).

The Branchial Formula of ARACHNODROMIA.

Somites and their appendages.	Podobranchiæ.	Arthrobranchiæ.	Pleurobranchiæ.		
VII.	0 ep.	0	0	=	ep.
VIII.	1+ep.	1	0	=	2+ep.
IX.	1+ep.	2	0	=	3+ep.
X.	1+ep.	2	0	=	3+ep.
XI.	1+ep.	2	1	=	4+ep.
XII.	1+ep.	2	1	=	4+ep.
XIII.	0	2	1	=	3
XIV.	0	0	1	=	1
	5+6 ep.	11	4		20+6 ep.

The formula is thus the same as that given by Bouvier for *Homolodromia*.

This crustacean so closely resembles the *Homolodromia* described and figured by Milne Edwards* and referred to by Bouvier,† that at first sight it might be supposed to be the same form.

In *Homolodromia*, however, it is distinctly stated that the antennules are not retractile, and that there are no special orbits.

In *Arachnodromia*, on the other hand, there are orbits formed on exactly the same plan as, and hardly less perfect than, those of *Dromia*, and they afford complete protection to the retracted eyes and antennules, the antennulary flagella folding, as in *Dromia*, behind the eyes.

* A. Milne Edwards, Bull. Mus. Comp. Zool. Vol. VIII. 1880, p. 32, and Recueil de figures de Crustacés Nouveaux etc. pl. 39, fig. 2. Not the Homolodromia of Miers.

† E. L. Bouvier, Bull. Soc. Philom. Paris (8) VIII. 1895–96, p. 37, *et seq.*

Arachnodromia Baffini, Alcock and Anderson, Plate II. fig. 1.

Carapace square-cut, dorsally convex, very distinctly (from a fourth to a fifth) longer than broad, its greatest breadth being just in front of the posterior border, its greatest depth approximating its greatest breadth, its surface—like that of the appendages and other parts of the body—tomentose. Except for a few small sharp granules anteriorly and laterally and along the lateral border, the carapace is unarmed.

The front is deeply cleft to its base, and has the form of two acutely triangular teeth.

Upper margin of orbit notched near its outer angle which is dentiform, the outer angle of the lower margin of the orbit is much more strongly dentiform, and the (outer) orbital wall between the two spines is deficient.

Antennal flagella longer than the carapace.

Chelipeds rather slender, unarmed except for a few granules seen on denudation, about $1\frac{2}{3}$ times the length of the carapace: fingers strongly hollowed '*en cuillère*,' especially the immovable one, which alone has teeth: wrist not elongate.

First two pairs of legs more than twice the length of the carapace: their dactyli are about two-thirds the length of the preceding joint, are stout, are sharply spinate along the posterior edge, and end in a claw. The last two pairs of legs are about the same length as the carapace: their small claw-like dactyli shut down on a ring of spines at the end of the preceding joint.

Colours: dirty whitish, with a bluish tinge on the carapace and a faint reddish tinge elsewhere; eyes chocolate.

Two males and a female, from off the Travancore coast, 430 fms.: a small male from the Andamans, 238-290 fms.

The carapace of the largest male is 20 millim. long and 15 millim. broad, that of the female is 30 millim. long and 24 millim. broad.

Named in memory of the great Arctic explorer William Baffin, who, according to Sir Clements Markham, was the first Englishman to actually plot charts in these Seas.

DYNOMENE, Latreille.

Dynomene, Latreille, in Cuvier's Règne Animal, (nouv. éd. 1829, p. 69): Desmarest, Consid. Gen. Crust. p. 133: Milne Edwards, Hist. Nat. Crust. II. 179: Lamarck, Hist. Nat. Anim. Sans. Vertebr. 2nd edit. p. 482: De Haan, Faun. Japon, Crust. p. 104: Dana, U. S. Expl. Exp. Crust. pt. I. p. 402: A. Milne Edwards, Ann. Sci. Nat. Zool. (6) VIII. 1879, Art. 3.

Dynomene margarita, n. sp. Plate II. fig. 3,

The whole carapace and dorsal surfaces of the chelipeds and legs are as closely as possible covered with prickly spines and spinules: the under-surfaces of the body and legs, eyestalks, antennæ and external maxillipeds are crisply granular. On the middle of the fourth abdominal tergum is a pair of large tubercles, just like pearls, in the closest contact with one another.

Carapace subglobular, longer than broad; the regions hardly indicated, though the cervical groove is fairly plain.

Front triangular, concave, deflexed, its apex continuous with that of the epistome. Supra-orbital borders tumid.

Antennal peduncle stout, the flagellum nearly as long as the carapace.

Chelipeds equal, a little longer and stouter than the first three pair of legs and not much longer than the carapace: the fingers are short and stout, and meet throughout their extent.

The last pair of legs are mere rudiments, being slender and hardly longer than the basal joints of the preceding pair.

Colours in spirit, milk white, eyes deeply pigmented.

A single small male from the Andaman Sea, 75 fathoms: the length of the carapace is 5 millim.

Except for its legs, only the last pair of which is reduced in size and is dorsal in position, this species, by reason of its subglobular carapace, looks far more like a *Dromia* than a *Dynomene*.

Brachyura Oxystoma.

Family *Calappidæ*.

Calappa, Fabr.

Calappa, Fabricius, Ent. Syst. Suppl. p. 345: Bosc, Hist. Nat. Crust. I. p. 179: Latreille, Hist. Nat. Crust. et Ins. V. p. 389: and in Cuvier's Règne Animal, 2nd ed. 1829, p. 66: Lamarck, Hist. Nat. Anim. Sans. Vertebr. V. p. 484: Desmarest, Consid. Gen. Crust. p. 108: Risso, Hist. Nat. Eur. Mérid. V. p. 30: Roux, Crust. Médit. p. 5: Milne Edwards, Hist. Nat. Crust. II. 102: De Haan, Faun. Japon. Crust. pp. 69, 70, 125: Dana, U. S. Expl. Exp. Crust. pt. I. p. 391: Heller, Crust. Sudl. Europ. p. 129: Miers, Challenger Brachyura, p. 283: Alcock, J. A. S. B. Vol. LXV. pt. 2, 1896, p. 139.

Carapace strongly convex, rounded in front, much broadened behind by a pair of clypeiform expansions, or wings, beneath which the four pairs of ambulatory legs are concealed in flexion.

Front small, somewhat triangular, projecting little or not at all beyond the level of the orbits, bilobed.

Orbits small, circular: eyestalks short and thick.

The antennules fold nearly vertically beneath the front.

The basal joint of the antennæ is very broad, and fills a wide hiatus at the inner angle of the orbit: the flagellum is short usually.

There is no distinct epistome; but the endostome is prolonged, as far as the antennulary fossæ, in the form of a canal, which is divided longitudinally by a deep vertical septum into two channels, each channel being completed below by a lamellar process from the first pair of maxillipeds.

The external maxillipeds do not meet across the mouth, but leave exposed between them the mandibles, and, in front of them, the afore-mentioned leaf-like prolongations from the first pair of maxillipeds.

The chelipeds are very large, and in flexion are closely apposed to the front half of the carapace, so as to form a sort of buckler: the arm has near its distal end, externally, a transverse wing-like expansion, complementary to the wing-like expansions of the carapace: the hand is strongly compressed, its upper border forming a high, sharply dentate or crenulate, crest. Except for the fingers, the chelipeds are equal and symmetrical; both the fingers, namely, of one hand have on their outer aspect, near the base, a stout projecting lobule.

The abdomen in the adult male consists of only five separate pieces, owing to the fusion of the 3rd, 4th and 5th somites. In the young male, as in the adult female, it consists of seven separate somites.

Calappa exanthematosa, Alcock and Anderson.

Calappa exanthematosa, Alcock and Anderson, J. A. S. B. Vol. LXIII. pt. 2, 1894, p. 177, and Ill. Zool. Investigator, Crust. pl. xv. figs. 1, 1a: Alcock, J. A. S. B. Vol. LXV. pt. 2, 1896, p. 146. ? = *Calappa japonica,* Ortmann, Zool. Jahrb. Syst., etc. VI. 1892, p. 566.

Extreme length of carapace a little more than two-thirds the extreme breadth.

The carapace is generally inflated, especially in the branchial regions: its surface in rather more than its anterior half is covered with large round, or oval, smooth mamillary tubercles having a red base and a shining yellow apex, and exactly resembling smallpox pustules; and is covered posteriorly with smaller round, or oval, slightly elevated patches, which exactly resemble smallpox papules. The antero-lateral borders of the carapace are quite smooth in their anterior half, and have 4 or 5 coarse serrations in their posterior half: the posterior border is beaded, and is bounded on either side by a tooth.

The clypeiform expansions are little developed, their extreme transverse dimension being less than one-third their extreme dimension in an inwardly oblique antero-posterior direction: they consist of about seven serrated teeth.

The pterygostomian regions have only a few scanty hairs.

The front is bifid, the breadth of its tip is half again that of the orbit, beyond which it does not project.

The flagellum of the antenna is nearly twice the breadth of the orbit in length.

The endostomial septum is narrow, not extending vertically to the level of the mouth, and quite plainly shows its origin out of a fold of the endostome: its anterior border is cut straight, and projects obliquely.

The wing-like expansion at the end of the arm has its edge finely serrate and 4-dentate. The upper surface of the wrist and the outer surface of the palm are more or less covered with pustules similar to those on the carapace. The palm has its crest sharply 6- or 7-dentate and its lower surface uniformly covered with bead-like granules.

The sterna corresponding to the 2nd, 3rd and 4th pairs of legs are much inflated.

Bay of Bengal, off the Madras coast, 91-112 fms.

In the *young* the tubercles on the carapace are sharper, and extend further backwards.

Ortmann, Zool. Jahrbucher, Syst. etc., X. 1897, p. 296, states that this species is absolutely identical with the *Calappa japonica* previously established by him in Zool. Jahrb. Syst. etc. VI. 1892, p. 566.

Mursia, Desmarest, Edw.

Mursia, Desmarest, Consid. Gen. Crust., p. 108, pl. 9, fig. 3: Latreille, in Cuvier, Règne Animal, ed. 2, 1829, p. 39; and Milne-Edwards in Cuvier, Règne Animal, ed. 3, p. 64: Milne-Edwards, Hist. Nat. Crust. II. 109: Lamarck, Hist. Nat. Anim. sans Vertebr. V. 486: De Haan, Faun. Japon. Crust. p. 68 and p. 125: Dana, U. S. Expl. Exp. Crust. pt. 1. p. 391: Miers, Challenger Brachyura, p. 290, (*ubi synon.*): Alcock, J. A. S. B. Vol. LXV. pt. 2. 1896, p. 149.

Thealia, Lucas, Ann. Soc. Entomol. France (I) VIII. 1839, p. 577.

Carapace oval, moderately convex, rounded in front, rather suddenly contracted behind, the evenly-arched antero-lateral margins ending in a large lateral epibranchial spine.

Front with a small acuminate tip.

Orbits rather large, oval, with at least one closed but distinct fissure in the upper margin, and with two wide gaps in the lower margin, in one of which the basal joint of the antenna is lodged. Eyes large, eyestalks short and thick.

The antennules fold obliquely. The basal joint of the antennæ is not dilated.

There is no distinct epistome, but, as in *Calappa*, the endostome is prolonged into a canal, which however is but incompletely divided longitudinally, the septum being little more than a ridge anteriorly, though well developed posteriorly. As in *Calappa* the first pair of maxillipeds give off each a lamellar process to complete this efferent canal below.

The external maxillipeds do not meet across the mouth, but, as in *Calappa*, leave exposed between them the mandibles, and, in front, the leaf-like prolongations of the first maxillipeds.

The chelipeds are enlarged, much as in *Calappa*; but the meropodite, or "arm," instead of a transverse crest near the distal end of its outer surface, has merely a ridge with one or two spines: the palm is compressed and its upper border forms a dentate crest, but not such a high one as that of *Calappa*. As in *Calappa* the chelipeds are only asymmetrical as regards the fingers, which on one hand have on their outer aspect, near the base, a stout lobule.* The legs are large, the first two pairs being at least as long as the chelipeds.

The abdomen in the male is as broad in the proximal half as it is in the female: in the adult male it consists of five segments, the 3rd, 4th and 5th being intimately fused, the sutures even being hardly distinguishable: in both sexes the tergum of the 1st somite is almost entirely concealed, and that of the 2nd somite strongly carinate transversely.

Mursia is practically *Calappa* without the wings to the carapace, and with large strong legs: the widely fissured orbital floor, the less pronounced endostomial septum, and the slender basal antenna-joint are the other important points of difference.

Mursia bicristimana, Alcock and Anderson. Plate III. fig. 3.

Mursia bicristimana, Alcock and Anderson, Journ. Asiatic Soc. Bengal, Vol. LXIII, 1894, pt. 2, p. 179; and Ill. Zool. 'Investigator' Crust. pl. xxiv. fig. 5: Alcock, J. A. S. B., Vol. LXV. pt. 2, 1896, p. 150.

The length of the carapace is about seven-ninths of the breadth immediately in front of the lateral epibranchial spine; and the length of the epibranchial spine is from one-third (in the young) to less than one-fourth (in the adult) the length of the carapace.

The surface of the carapace is closely granular, and in addition there are *seven rows of tubercles*, one in the middle line, and three on each side radiating over the branchial regions: the antero-lateral margins are finely beaded and evenly and sharply festooned: the postero-lateral margins are without the angular bend inwards seen in M. armata: *the posterior margin is bounded on either side by a laminar denticle, not by a great projecting lobule as in* M. armata.

The outer parts of the pterygostomian and subhepatic regions are covered with a dense felt of long hairs.

The rostrum is trilobed, its breadth at the level of the lobes being about one-half more than the greatest breadth of the orbit.

The transverse ridge near the distal end of the arm is very hairy, and is armed distally with two spines, the outer and larger of which is more than half the length of the lateral epibranchial spine. *This ridge is continued along the palm as a sharp longitudinal crest* (more prominent even than that of *Platymera*) which is unevenly trilobed, the proximal lobe being spiniform, the middle lobe broad and obtuse, and the distal lobe narrow and obtuse. The upper surface of the wrist, and the outer surface of the palm and fingers, are closely and sharply granular: the upper edge, or crest, of the palm is 7-serrate.

* In *Mursia hawaiiensis*, Mary J. Rathbun, Proc. United States National Museum, xvi. 1893, p. 252, the chelipeds are described as very unequal.

The ambulatory legs are large stout and compressed, those of the first three pairs being a little longer than the chelipeds. In these three pairs the *meropodite is lamellar, its greatest breadth being considerably more than a third its length*: the carpus has its outer surface traversed longitudinally by three beaded carinae, the middle one of which ends in a spine; and the propodite is lamellar with the outer (anterior) edge subcarinate and the upper surface traversed longitudinally by two or three raised lines of fine beading.

The second abdominal tergum in both sexes is raised into a stout carina, the height of which is more than a third the transverse diameter of the tergum: this carina is three lobed, the lobes being separated only by fissures. In the female, as in the male, the 3rd–5th terga are fused, although the lines of fusion are quite distinct in the former sex.

Colours in life salmon pink.

Off Ceylon, 142–400 fms., and 180–217 fms.

In the form of the legs, in the ornamentation of the chelipeds, and in the shape of the carapace, this species bears a strong resemblance to *Platymera*. Even in the articulation of the flagellum with the merus of the external maxillipeds the appearances are somewhat those of *Platymera*.

On the other hand the form of the endostomial channels, and of the processes of the first maxillipeds which close those channels ventrally, as well as the practical symmetry of the chelipeds, are all as in *Mursia*.

But a comparison of this species with specimens of *Mursia armata* and *Platymera gaudichaudii* leads to the belief that all three are congeneric.

The dimensions of an adult male are as follows:—
breadth of carapace 67 millim., excluding the lateral epibranchial spines; length of carapace 47 millim., length of first pair of ambulatory legs about 90 millim.

Mursia aspera, n. sp.

Illustrations of the Zoology of the Investigator, Crustacea, pl. xl. fig. 2. (in *preparation*.)

Most nearly allied to *Mursia curtispina*, Miers and to *Mursia hawaiiensis* Rathbun.

Carapace strongly convex in both directions, its length four-fifths of the breadth in front of the lateral epibranchial spine, its surface closely covered dorsally with sharp pustulous granules and tubercles, seven longitudinal series of which are slightly enlarged and have a granular base. *The lateral epibranchial spine is extremely short, its length in the unique* (adult male) *specimen being only one-twentieth the greatest breadth of the carapace.*

Rostrum trilobulate: antero-lateral borders finely-festooned, crenulate: postero-lateral borders beaded, angularly bent inwards about the middle as in

M. armata: posterior border with three sharp denticles of equal size—one at either end, one in the middle.

Pterygostomian regions channelled as usual, the surface outside the channels hairy.

Inner border of ischium of external maxillipeds deeply cut into elegantly-rounded, specially calcified teeth.

Chelipeds perfectly equal, the exposed outer surfaces closely covered with sharp granules and acutely-conical tubercles. Transverse crest at the far end of the arm with four spines, the outermost of which is the largest, but is of no great size. The only crest on the hand is that of the 8 or 9-toothed upper border.

Legs almost as in the preceding species, but shorter—the first three pair being hardly as long as the chelipeds—and having the dorsal surfaces of the meropodite and carpus more granular.

The second abdominal tergum has a very prominent transverse three-lobed carina, as in the preceding species.

A single adult male, from off the Maldives, 210 fathoms, has the carapace 48 millim. long and 60 millim. broad, and the lateral epibranchial spine 3 millim. long.

The specimen on which this species is founded was received during my absence in England, and was accidentally put away among the general collection, where it escaped notice until after the printing of this report was considerably advanced.

Family *Leucosiidæ.*

RANDALLIA, Stimpson.

Randallia, Stimpson, Journ. Boston Soc. Nat. Hist. Vol. VI. 1857, p. 471: Miers, Challenger Brachyura, p. 316: Alcock, J. A. S. B. Vol. LXV. pt. 2, 1896, p 191.

Carapace circular and convex, almost globular; with the front narrow, usually broadly bidentate, and somewhat sunk behind the level of the front edge of the buccal cavern. The subhepatic or pterygostomian regions are convex and puffed out, so as to increase the squat and sunken appearance of the front. There is a remarkably broad vertical interval between the orbits and the edge of the buccal cavern.

The surface of the carapace is, typically, covered with vesicular or pustulous granules, but these are sometimes visible only with a lens: the regions are usually, but not always, distinctly demarcated by grooves.

The posterior margin is generally, but not always, armed with spines or petaloid lobules or tubercles.

The orbits are somewhat imperfect; their upper edge is deeply emarginate, there is a wide gap at the inner canthus, and there are three very distinct sutures, or sometimes actual fissures, in the upper-outer wall.

The antennules fold obliquely: in one Indian species their basal joint forms a close-fitting operculum to the antennulary fossa. The antennæ are very distinct, and are loosely lodged in the inner canthus of the orbits.

The buccal cavern is triangular and somewhat elongate: the exognath is not dilated and its outer margin is almost straight: the triangular merus of the endognath is about $\frac{2}{3}$ the length of the ischium measured along its inner edge.

Chelipeds either massive or moderately stout, of moderate length; fingers stout, about as long as the palm, which is not more—but is usually much less—than half the length of the carapace.

Although there is, as usual, some fusion among the abdominal terga, yet the sutures are never wholly obliterated as they are in most other Leucosines.

Randallia lamellidentata, Wood-Mason.

Randallia lamellidentata, Wood-Mason, Illustrations of the Zoology of the 'Investigator' Crustacea, pl. v. figs. 5, 5a, 5b: Alcock, Ann. Mag. Nat. Hist. May, 1894, p. 404, and J. A. S. B., Vol. LXV. pt. 2, 1896, p. 195.

Carapace rhomboidal with the angles rounded off—subcircular; its surface behind the front covered with unequal-sized rather scattered pustulous tubercles; its regions well defined by grooves of some depth.

Front bluntly bidentate. On the antero-lateral margin are three broad lamelliform teeth, the front one of which is on the pterygostomian ridge (which as usual forms the front part of the antero-lateral margin), and there is a fourth similar tooth at the junction of the antero-lateral and postero-lateral margins. The postero-lateral margins are full and the pustulous tubercles extend on to them.

The short posterior margin is elegantly bilobed, with a few pearly granules round the lobes, and is overhung by the tip of the horizontal spine in which the intestinal region culminates.

The ventral surface of the carapace, the thoracic sterna, abdominal terga (in the male) and external maxillipeds are all granular, the granules above the base of the chelipeds being enlarged and pearly.

The chelipeds in the male are about two-thirds as long again as the carapace, and are massive and granular: at the distal end of the outer edge of the somewhat trigonal arm the granules are enlarged and almost spiniform, as are also one or two at the distal end of the outer surface of the wrist. The hand is not much longer than broad and hardly one-third the length of the carapace; its outer edge is in the form of a remarkably thin and deep crest: the fingers are stout and rather longer than the hand, their outer (non-opposed) edges are cristiform.

The legs are granular, the granules on the dorsum of the propodites carpopodites and distal end of the meropodites being spiniform, as also on the outer surface of the ischium and merus of the last pair: the dactyli are hairy.

The 3rd–6th abdominal terga of the male are fused but are all very distinctly and independently recognizable, the 6th has a terminal denticle.

The largest male, dredged in the Andaman Sea at 350 fms., has the carapace between 16 and 17 millim. long and 18 millim. broad (without spines).

Randallia pustulosa, Wood-Mason.

Randallia pustulosa, Wood-Mason, Ann. Mag. Nat. Hist. March, 1891, pp. 266 and 267, and Illustrations of the Zoology of the 'Investigator' Crustacea, pl. v. fig. 4: Alcock, J. A. S. B. Vol. LXV. 1896, pt. 2, p. 196.

Carapace subcircular, subspherical; covered with unequally large pustulous tubercles the surface of which, like the surface between them, is finely and closely granular under the lens; all the regions are well defined by broad grooves.

The front is narrow and is broadly bidentate. The lateral margins are full and inflated, and carry in the adult a series of tubercles, in the young a series of blunt spines: in the antero-lateral margin, between the hepatic and branchial regions, is a conspicuous notch, which corresponds with a groove or depression in the pterygostomian face of the carapace.

The short posterior border has a spine or dentiform lobe at either end, and is overhung by the long spine in which the tumid intestinal region culminates.

The whole under surface is densely granular in the young male, but in the female the fused 4th–6th abdominal terga and the inner half of the ischium of the external maxillipeds are smooth.

The chelipeds in the adult female and young male (adult male unknown) are twice the length of the carapace and are everywhere finely granular. The hand is subcylindrical and elongate, being half as long as the carapace; the fingers are stout and about as long as the hand, they are finely denticulate, with enlarged denticles at regular distant intervals.

The legs are stoutish and, to the naked eye, smooth: the dactyli are fringed with hairs.

In the (young) male the 3rd–6th abdominal terga are fused but without any obliteration of sutures: in the adult female the 4th–6th are fused and the sutures obliterated.

Carapace of an adult female about 31 millim. in either diameter.

In the adult female the brood-pouch communicates with the branchial chambers on either side by means of a foramen, as in *Parilia*.

From the Andaman Sea, 240–220 fms., and 250 fms., and from the Laccadive Sea, off the Travancore coast, 406 fms.

In the young the carapace is quite spherical, with its edges spiny and its surface closely and crisply granular.

PARILIA, Wood-Mason.

Parilia, Wood-Mason, Ann. Mag. Nat. Hist., March 1891, p. 264: Alcock, J. A. S. B. Vol. LXV. pt. 2, 1896, p. 198.

Carapace strongly convex, especially posteriorly, somewhat oval transversely, with three spines on the posterior margin; the surface finely granular, the regions fairly well-defined.

The front is narrow and bidentate, and the epistome projects well beyond it, the epistome being, for an Oxystome, deep—as in *Randallia* and *Nucia*.

The eyes are small, and the orbits imperfect, for not only have they two fissures (not mere sutures) in the roof, and a broad fissure in the outer wall, and a broad gap communicating with the antennary and antennulary fossæ, but their upper-outer wall is deeply emarginate.

The antennules fold a little obliquely. The antennæ are distinct, and stand in the gap at the inner canthus of the orbit, which they do not nearly fill.

The buccal cavern is considerably broader than long, owing to the enormous width of the afferent branchial channels and of the foliaceous expansion of the exognath that covers them: the outer edge of the latter is strongly curved: the triangular merus of the endognath is very nearly as long as the ischium, measured along the inner edge.

The chelipeds in the adult male are several times the length of the carapace, and are slender, though more massive than the legs: the hands are several times the length of the stoutish fingers.

The abdomen in the male consists of five distinct pieces: in the female it consists of seven, but the 4th, 5th and 6th are not separately movable.

Branchial chambers greatly inflated, especially posteriorly: branchiæ large, and six in number on either side. [Brood-pouch of the female very large and communicating with the branchial chamber on either side, at base, by a foramen.]

Parilia alcockii, Wood-Mason. Plate IV. fig. 1. ♂

Parilia alcockii, Wood-Mason, Ann. Mag. Nat. Hist., March 1891, p. 264, and Ill. Zool. 'Investigator,' Crust. pl. v. figs. 3, 3a ♂: Alcock and Anderson, J. A. S. B. Vol. LXIII. pt. 2, 1894, p. 177: Alcock, J. A. S. B. Vol. LXV. pt. 2, 1896, p. 198.

Carapace about seven-eights as long as broad, transversely oval, but with the anterior margin—between the outer angles of the afferent branchial channels—perfectly straight.

The antero-lateral margin is broadly indented at the junction of the hepatic and branchial regions, and bears four denticles; and there are three denticles on the posterior margin, the middle one of which is the smallest: just above the posterior margin is another transverse row of three denticles,—one in the middle of the intestinal region and one on the posterior wall of the branchial region on either side.

The carapace is strongly convex, the convexity gradually increasing from before backwards and then suddenly dropping, like a simian cranium, which in profile it much resembles: the surface is everywhere finely granular.

The regions of the carapace are well delimited by broad shallow grooves and lines of dimples, the branchial regions each forming an enormous tumid expanse. A slightly raised ridge traverses the carapace, in the middle line, from the base of the front to the intestinal denticle.

The front is broadly bilobed, each lobe being convex dorsally and acuminate: beyond it in a dorsal view is seen the epistome and the whole length of the edge of the buccal cavern.

The surface of the external maxillipeds and the ventral surface of the carapace are finely granular, but the sternum and the greater part of the abdomen are smooth. In the middle of the sternum of the female, between the genital openings, is an erect spine.

The external maxillipeds have a narrow triangular endopodite, the merus of which is strongly curved upwards towards the front; and a foliaceous exopodite, which is much shorter than the endopodite, and which is semicircular in shape and two-thirds as broad as long—broader even than in *Philyra globosa*, Fabr.

The chelipeds as in *Myra fugax*, vary according to age and sex: in the adult male they are $4\frac{1}{2}$ times, in the female and young male $2\frac{1}{2}$ times, the length of the carapace, and are only about twice as massive as the legs: their surface up to nearly the end of the hand is finely scabrous. The arm is cylindrical: the hand in the female is cylindrical, but in the male somewhat clavate. The hand in the male is more than 3 times, in the female only twice the length of the fingers: the fingers are stout, gently curved in the female, somewhat sinuous in the adult male, and their opposed edges are almost edentulous.

The legs in the male are shorter than the arm; in the female they are a little longer than the arm: they are cylindrical, and finely scabrous on the dorsal surface: the dactyli are obtusely pointed, and have both their edges closely fringed with longish stiff hairs.

Colours in spirit rusty reddish.

The carapace of the average adult male is 50 millim. long and 56 millim. broad, of the adult female 40 millim. long and 48 millim. broad.

Fairly common on soft muddy bottoms along the east coast of India between 70 and 250 fathoms.

In the Indian Museum collection are 96 specimens of both sexes and all ages.

The figure of this fine species exhibited on plate V. of the *Illustrations of the Zoology of the Investigator* is a female. The adult male is here figured.

Pariphiculus, Alcock.

Pariphiculus, Alcock, J. A. S. B., Vol. LXV. pt. 2, 1896, p. 257.

Closely allied to *Iphiculus*, but differing in several important characters and in the whole form of the carapace. The appendages are as densely tomentose as in *Iphiculus*, but the carapace is covered with a finer and sparser tomentum which does not quite conceal the texture of the surface.

The carapace is circular and globular, with its margins coarsely spinate, and its surface vesiculous: the intestinal region is very distinctly isolated, but the other regions are almost lost in the general convexity of the carapace.

The front is narrow: in one species it projects as a distinct snout, in the other the angle of the afferent branchial canal can be seen beyond it in a dorsal view, but the whole mouth can never be seen beyond it as it can in *Iphiculus*.

The orbits are obliquely elongate and completely conceal the eyes: two distinct fissures are plainly visible in the emarginate roof, besides a fissure in the lower part, and there is a gap at the inner canthus where the basal joint of the antenna—the flagellum of which is large—stands. The antennules fold very obliquely. There is a space of varying width between the edge of the orbit and the edge of the buccal cavern.

The buccal cavern is rather elongate-triangular, and the merus of the external maxillipeds is half the length of the ischium measured along the inner border.

The chelipeds are from $1\frac{1}{4}$ to $1\frac{2}{3}$ times the length of the carapace: the hand is short, cylindrical with the base inflated, or is subglobular, but not nearly so swollen as in *Iphiculus*: the fingers are slender, much longer than the hand and somewhat hooked; they open in an obliquely vertical plane, and the tip of the mobile finger moves through the usual arc of about 75°. The legs are moderately stout.

The abdomen of the male has the 3rd, 4th and 5th segments fused: that of the female has all the segments distinct.

Pariphiculus coronatus, Alcock & Anderson.

Randallia coronata, Alcock & Anderson, J. A. S. B., Vol. LXIII. pt. 2, 1894, p. 177.

Pariphiculus coronatus, Alcock Ill. Zool. 'Investigator,' Crust. pl. xxiv. fig. 2: and J. A. S. B., Vol. LXV. pt. 2, 1896, p. 258.

Carapace globular, just broader than long, its surface closely covered with large vesiculous granules beneath a dense fine-textured pubescence: the intestinal region forms an independent circular swelling, bounded by a deepish groove and surmounted by two spiniform tubercles, one behind the other: the gastric region is partly defined anteriorly by two creases, and the cardiac region is partly defined posteriorly by two grooves, and a narrow and indistinct groove separates the hepatic from the branchial region on either side: on either lateral margin are 5 spiniform tubercles, not including the dentiform prolongation of the outer angle of the buccal cavern, and at either end of the short posterior margin is a dentiform tubercle: 3 similar tubercles occur, one in the middle of the cardiac region and one on either side of it on the after part of the branchial regions—these three, along with the last on the lateral borders and the two on the posterior margin, forming a ring round the tumid intestinal region: the side-wall of the carapace is grooved longitudinally just above the epimeral edge.

The front is bidentate, its tips just projecting beyond the level of the buccal cavern.

The chelipeds in the female (male unknown) are $1\frac{2}{3}$ times the length of the carapace: the hand is inflated, cylindrical, and about $\frac{3}{4}$ the length of the fingers: the fingers are very slender, almost hairless, hooked at tip, finely denticulate with a few slightly larger denticles at distant intervals, and they open in an obliquely vertical plane.

Length of carapace of female (apparently adult) 16 millim., breadth 17 millim.

From the Bay of Bengal, off Coromandel coast, 112 fms.

Family ***Dorippidæ.***

Ethusa, Roux.

Ethusa, Roux, Crust. de la Méditerranée, pl. xviii, and text relating thereto: Milne Edwards, Hist. Nat. Crust. II. 161: Lamarck, Hist. Nat. Anim. sans Vertebr. (2nd ed.) Vol. V. p. 417.

Ethusina, S. I. Smith, 'Albatross' Crustacea, 1883, in Ann. Rep. U. S. Comm. Fish, &c., 1882 (1884).

Ethusa, Miers, 'Challenger' Brachyura, pp. 328, 331: Alcock, J. A. S. B., Vol. LXV, pt. 2, 1896, p. 281

Carapace shaped much as in *Dorippe*. The front consists of two laminar teeth each of which again is bifid at tip: on either side of the front, and separated from it by a deep cleft, is a long flat tooth or spine formed by the prolongation of the antero-external angle of the carapace, and forming the outer angle of the orbit. There is practically no orbital floor. The antennules fold obliquely: they are large, but fold fairly well into their fossæ. The antennæ have a long flagellum: their basal joint is inserted between the eyestalk and the basal antennulary joint, but on a slightly lower level.

The buccal cavern is elongate-triangular and does not usually extend to the front: the external maxillipeds cover only its basal three-fourths, or thereabout, somewhat as in *Dorippe*, but the distal part is closed in by stout foliaceous processes of the first maxillipeds. The flagellum or palp of the external maxillipeds arises near the antero-external angle of the rather broad merus, and is completely exposed in flexion.

The afferent branchial orifices are wide openings immediately in front of the bases of the chelipeds.

The chelipeds in the adult male are often unequal: the legs have the same form and relations as in *Dorippe*, but the last two small and dorsally placed pairs are not subchelate, although their little hook-like dactylus folds backwards. The dactyli of the 1st and 2nd pairs are palmulate and are very long and stout.

The abdomen of the male usually consists of 5 pieces, the 3rd–5th terga being fused, that of the female consists of 7 separate terga. As in *Dorippe* the first three terga are visible in a dorsal view.

There is very little hair about the carapace and larger appendages.

In the Indian seas the species of this genus are, so far as is known, found only at depths of between 200 and 1,300 fathoms.

Ethusa indica, Alcock.

Ethusa indica, Alcock, Ann. Mag. Nat. Hist., May, 1894, p. 405, and Ill. Zool. 'Investigator,' Crust. pl. xiv. fig. 2. ♀, and J. A. S. B. Vol. LXV. pt. 2, 1896, p. 283.

Carapace convex; its extreme length, including the frontal teeth, in the male only just exceeds, and in the female equals, its extreme breadth; its surface is finely and closely granular almost everywhere, except sometimes on the cardiac-intestinal region.

The branchial regions are much swollen, both dorsally and laterally, the lateral swelling making the carapace more than one-third broader across the middle of the branchial regions than across the bases of the external orbital spines. The cardiac-intestinal region is small and well defined, and although it is tumid it is commonly sunk below the level of the branchial convexities. The anterior regions of the carapace are undefined.

The spine at the external orbital angle is broad-based, but long, slender, and acute: it projects obliquely outwards well beyond the tips of the frontal teeth. The two pairs of frontal teeth are longish and acute—the outer pair being somewhat the longer: they as well as the external orbital spine are a good deal concealed in a fringe of long hairs.

The eyestalks are short slender and freely movable: the eyes are often a little deficient in pigment.

The basal antennule-joint is not abnormally enlarged.

The chelipeds in the adult male only are asymmetrical, all the joints of one side being enlarged in all dimensions: the smaller cheliped is hardly as stout as the first two pairs of legs.

The second pair of true legs are not very much longer than the first: in the adult male they are a little more than three times the length of the carapace, and slightly more than three times the length of the 4th (last) pair; in the female they are not quite three times the length of the carapace, and about $2\frac{3}{4}$ times the length of the 4th pair.

The abdomen of the male consists of 5 pieces, the 3rd–5th terga being fused together.

The extreme length of the carapace is in the fully adult male 16·5 millim., in the fully adult female 15 millim.; the breadth 16 millim. in the male, 15 millim. in the female.

Has been dredged in the Andaman Sea at 240 fms., in the "Swatch" of the Gangetic Delta at 409 and at 405 to 285 fms., in the Laccadive Sea at 360 and 696 fms., off the Maldives at 719 fms., and off both coasts of Ceylon at 406 to 296 fms.

Ethusa pygmæa, Alcock.

Ethusa pygmæa, Alcock, Ann. Mag. Nat. Hist., May 1894, p. 406, and Ill. Zool. 'Investigator,' Crust. pl. xiv. fig. 5, ♀, and J. A. S. B., Vol. LXV. pt. 2, 1896, p. 284.

Distinguished from *E. indica* only in the following particulars:—

(1) its size is much smaller, the largest known specimen—an ovigerous female—having the carapace slightly over 6 millim. long and nearly 7 millim. broad:

(2) the external orbital spines, though of the same slender acute shape, are not so prominent, not reaching as far as the tips of the frontal teeth:

(3) the anterior regions of the carapace are plainly defined by grooves.

Andaman Sea 188 to 220 fathoms, and 240 to 220 fms.

Ethusa andamanica, Alcock.

Ethusa andamanica, Alcock, Ann. Mag. Nat. Hist., May 1894, p. 405, and Ill. Zool. 'Investigator,' Crust. pl. xiv. fig. 8, young ♀, and J. A. S. B., Vol. LXV. pt. 2, 1896, p. 284.

Carapace flat, its extreme length only just exceeds its extreme breadth, its surface finely granular under the lens, but smooth to the naked eye.

The branchial regions are a little tumid dorsally, but do not bulge laterally, so that the convergent lateral borders are nearly straight.

The external orbital spine is broadly triangular, with a mucronate tip which does not quite reach to the tips of the frontal spines; these also are acutely triangular, and all are a good deal hidden by a fringe of long hairs.

The eyestalks are short and rather stout, movable, but not very freely so: the eyes are not deficient in pigment. The basal antennule-joint is not enlarged.

The chelipeds of the adult male are unknown: in the female they are not so stout as the first two pairs of legs.

The second pair of legs in the female (adult male unknown) exceed the first almost by the length of the dactylus, they are three times the length of the carapace and about $2\frac{1}{2}$ times the length of the 4th pair.

The extreme length of the carapace of the largest specimen, which is not adult, is 9·5 millim., the extreme breadth 9 millim.

Andaman Sea 188 to 220 fms., and 238 to 290 fms.

This species may possibly be only a variety of *Ethusa orientalis*, Miers. Challenger Brachyura, p. 330, pl. xxviii. fig. 4.

Subgenus Ethusina, S. I. Smith.

In *Ethusina* the carapace has a slightly more elongate appearance, and the basal joint of the antennules is enlarged swollen and globular so as to push the eyestalks permanently outwards.

Ethusa (Ethusina) gracilipes, Miers.

Ethusa (Ethusina) gracilipes, Miers, Challenger Brachyura, pp. 332, 333, pl. xxix.
Ethusina gracilipes, Faxon, Albatross Stalk-eyed Crustacea, Mem. Mus. Comp. Zool. Vol. XVIII. 1895, p. 36.

Of this species, which is known from deep water in the East Indian Archipelago and also off the Pacific coast of Central America, the Investigator has dredged one specimen in 836 fms., between the Maldives and C. Comorin.

Ethusa (Ethusina) investigatoris, Alcock.

Ethusa (Ethusina) investigatoris, Alcock, J. A. S. B. Vol. LXV. pt. 2. 1896, p. 285.

Carapace manifestly longer than broad, somewhat convex, smooth to the naked eye though finely granular under the lens.

The branchial regions are a good deal swollen both dorsally and laterally, bulging out the lateral margins and making the carapace a third broader across the middle of the branchial regions than across the bases of the external orbital spines.

The cardiac-intestinal region is well-defined and tumid, but not sunk below the level of the branchial convexities: the anterior regions of the carapace are fairly well defined.

The frontal portion of the carapace is separated from the rest of the carapace by a transverse groove or crease. The external orbital spine is long and needle-like, but its tip falls considerably short of the tips of the rather long acute frontal spines.

The basal antenna-joint is huge and swollen, almost globular in shape. Owing to its size the eyes are pushed outwards until the eyestalks have come to lie almost in the tranverse axis of the carapace, with the tips of the eyes just visible, dorsally, beyond the lateral edge of the external orbital spine; and in this position they are almost immovably fixed.

The chelipeds in the apparently adult male are symmetrical and are not much stouter, except as to the hands, than the first two pairs of legs; the hands, however, are somewhat enlarged.

The second pair of true legs exceeds the first by about a third of the length of the dactylus; they are more than three times the length of the carapace, and about $2\frac{3}{4}$ times the length of the 4th pair.

The abdomen of the male consists of 5 pieces, the 3rd–5th terga being fused together.

Length of carapace of an adult male 12·3 millim., extreme breadth 11·3 millim.

Colours in life milk-white with the tip of the legs faint pink.

From the Bay of Bengal 1,300 fms. and Laccadive Sea 1,200 fms.

This species differs from *E. gracilipes*, Miers (1) in having the frontal portion marked off from the rest of the carapace by a transverse groove, and (2) in the different proportions of the eyestalks, which are long enough to expose the whole of the eye beyond the sides of the external orbital spines, in a dorsal view.

Ethusa (Ethusina) desciscens, Alcock.

Ethusa (Ethusina) desciscens, Alcock, J. A. S. B., Vol. LXV. pt. 2, 1896, p. 286.

This species is the connecting-link between *Ethusa* and *Ethusina*, and justifies the union of the two genera, proposed by Miers; for although it has the huge globular basal antennule-joint, the eyestalks retain a certain amount of mobility.

It closely resembles *E. gracilipes* and *E. investigatoris*, only differing from the latter (1) in its smaller size, (2) in having the hand of one cheliped—in the male—much larger than the other, and (3) in the greater mobility of the eye-stalks. The eyestalks moreover are short, like those of *E. gracilipes*, and do not expose the eyes to dorsal view on either side of the external orbital spine of the carapace.

From the Andaman Sea, 265 fms., and Laccadive Sea 912–931 fms.

Cymonomops, Alcock.

Cymonomops, Alcock, Ann. Mag. Nat. Hist., May 1894, p. 406, and J. A. S. B. Vol. LXV. pt. 2, 1896, p. 286.

Carapace of the *Dorippe* type (that is to say having its greatest breadth at its extreme posterior limit and leaving about half of the abdominal terga exposed to dorsal view), but arched anteriorly almost in a semicircle; its regions well defined in much the same way as *Dorippe*. The front is narrow and the whole fronto-orbital region lies well inside the semicircular curve of the antero-lateral margins: the narrow front ends in two little teeth between and beyond which can be seen the roof of the greatly prolonged buccal cavern, as in *Dorippe polita*. On either side of the front is a spine that forms the roof of the orbit, and outside of this spine, and separated from it by a deep notch, is a spine that forms the outer wall of the orbit.

The eyestalks are slender, moderately long, and freely movable: the eyes are almost without pigment.

The antennules have their basal joint lodged in a deep crevice between the edge of the anterior prolongation of the buccal cavern and the antennæ: their long flagellum cannot be concealed in flexion. The antennæ are large, but are much smaller than the antennules.

The buccal cavern is of great size,—not much less than half the length of the body, and is gradually narrowed anteriorly, and prolonged beyond the tip of the front: it is closed, except at its extreme frontal tip, by the long narrow external maxillipeds, the merus of which is not very much shorter than the ischium measured along the inner border and the flagellum of which is exposed in flexion: the long narrow pointed exognath is not much longer than the ischium: beneath the external maxillipeds the anterior prolongation of the buccal cavern is closed in below by a lamellar process of the first maxillipeds.

The chelipeds in both sexes are short, massive, and equal and symmetrical: the hands are of the chopper-shaped, almost subcheliform, Raninoid type, the stout fingers being almost at right angles to the long axis of the hand.

The first and second pairs of true legs are stout and are of great length, their merus being of relatively enormous length: the third and fourth pairs on the other hand, which are dorsal in position as in *Dorippe*, are extremely short and of filiform tenuity.

The abdomen in both sexes consists of six segments: in the male two or three of them are fused and the whole abdomen is very small, in the female the last segment is of great size.

[? The afferent branchial opening appears to lie in the deep crevice between the base of the antennæ and the edge of the buccal frame, in which the basal joint of the antennules is lodged.]

Cymonomops glaucomma, Alcock.

Cymonomops glaucomma, Alcock, Ann. Mag. Nat. Hist., May 1894, p. 406, and Ill. Zool. 'Investigator,' Crustacea, pl. xiv. fig. 9, and J. A. S. B. Vol. LXV. pt. 2, 1896, p. 287.

Carapace subcircular; it and the appendages are very closely and finely granular beneath a dense pubescence. The front consists of three deeply cut lobes, the middle one of which is the true front and is the largest and most prominent. The middle lobe again is slightly cleft at the tip, and in the cleft is to be seen projecting the roof of the remarkably prolonged buccal cavity.

The external orbital angle, which is somewhat ventrad in position, also forms a projecting tooth, so that the orbito-frontal region, which is sharply delimited from the rest of the inflated carapace, has the form of a five-pronged crest or crown. The regions of the carapace are plainly delimited, excepting only in the case of the boundary between the gastric and cardiac regions. The pterygostomian regions are most remarkably puffed out.

The abdomen in the female is large, and the terminal segment has the form of a broad semicircular plate, broader than any of the other segments and nearly as long as all of them put together: in the male the abdomen is very small.

The orbits are capacious, but the eyestalks are slender and the eyes are unpigmented and semi-opaque.

The antennules, which are much larger and longer than the antennæ, are incapable of flexion beneath the front.

The external maxillipeds are of great length, in correspondence with the remarkable trough-like prolongation of the buccal cavity, which they completely close in below; their meropodite, which is prolonged far beyond the insertion of the palp, covers the bases of the antennules and antennæ, their tips in fact being visible from above; the slender exopodite does not much surpass the ischium.

The chelipeds are short but massive, and are equal: the arm is curved, the wrist is small, the palm is large and tumid, and the fingers, which are set almost at right angles to the hand, are broad, compressed, pointed, very closely apposable, and have their cutting-edge very finely denticulated.

The first two pairs of legs are of great length, being more than four times the length of the body, the merus forming more than half their extent; their dactylus is filiform and is not much longer than their propodite. The last two pairs of legs have the family position, but are mere rudiments, being of hair-like tenuity and only about three-fourths of the carapace in length; the last pair ends in a hook-like dactylus.

A female from the Andaman Sea, 405 fathoms, has the following dimensions:—Length of carapace 6·5 millim., breadth 6·5 millim., length of cheliped 9 millim., length of second leg 28·5 millim., of fourth leg 4·5 millim. A male from the Andaman Sea, 265 fathoms, is smaller.

Colour in the fresh state chalky pink.

Family *Raninidæ*.

LYREIDUS, De Haan.

Lyreidus, De Haan, Faun. Japon. Crust. p. 138: Dana, U. S. Expl. Exp. Crust. pt. 1. p. 404: Haswell, Cat. Austral. Crust. p 114: Henderson, Challenger Anomura, p. 33: Alcock, J. A. S. B. Vol. LXV. pt. 2, 1896, p. 294.

Carapace elongate-obovate, the antero-lateral margins independent and gradually convergent; strongly convex from side to side and slightly convex from before backwards; smooth and polished, with the regions undefined. Fronto-orbital border less than half the breadth of the carapace. Eyes small; eyestalks short, broad at base, orbits hardly oblique.

Antennules about equal in size to the antennæ: antennæ with a stoutish peduncle and rather short slender flagellum, the peduncle not concealing the antennulary peduncle.

Merus of the external maxillipeds a little longer than the ischium.

Sternum broad as far as the bases of the first pair of true legs, then becoming narrow. Last pair of legs abnormally short and slender, arising well in advance of the penultimate pair. The abdomen in both sexes consists of 7 distinct segments.

Lyreidus Channeri, Wood-Mason.

Lyreidus channeri, Wood-Mason, P. A. S. B., August, 1885, p. 104, and J. A. S. B., Vol. LVI. 1887, pt. 2. p. 206, pl. i: Alcock, J. A. S. B. Vol. LXV. pt. 2, 1896, p. 294.
Lyreidus gracilis, Wood-Mason, J. A. S. B., Vol. LVI. 1887, pt. 2, p. 376.

The greatest breadth of the carapace—considerably in rear of the front—is a good deal more than half its greatest length, and is about $2\frac{1}{2}$ times the width of the fronto-orbital border.

The rostrum consists of a simple flat acutely-triangular spine; on either side of it, projecting beyond it, separated from it by a deep bight, and parallel with its tip, is a long acicular spine forming the external orbital angle. The fronto-orbital region is hairy.

The gradually convergent antero-lateral borders are about two-fifths the length of the postero-lateral borders, the junction of the two borders being occupied by a long oblique acicular spine; and nearly midway between this spine and the spine at the external angle of the orbit on either side, is another similar but rather shorter spine. The postero-lateral borders are defined in more than their posterior half by a very fine raised line.

The surface of the carapace is finely and closely punctulate in all its anterior half, as are also the pterygostomian regions.

The eyestalks are broad and flat, and taper to the cornea, which has a somewhat lateral position and is a little deficient in pigment. The arms have a spine or two little spines near the middle of their dorsal surface: the wrist has a large spine in the distal half of its upper border: the hand has its outer (upper) edge carinate up to a subterminal denticle, and has its lower edge cut into two or three sharp teeth: the dactylus has its cutting edge faintly and irregularly sinuous, but by no means denticulate, and the opposed edge of the immobile finger is irregularly and rather bluntly jagged. The legs are almost free from hair, a few hairs occurring on the posterior edge of the propodite and dactylus of the third pair and on the last two joints of the rudimentary fourth pair only: in the first and third pairs the carpus is dorsally carinate and the propodite foliaceously expanded, in the first and second pairs the dactylus is little more than broadly palmulate, and in the third pair the dactylus is foliaceous. The third and fourth abdominal terga are armed each with a median recurved spine, in both sexes.

The largest female in the Indian Museum collection has the carapace 28·5 millim. long, a smaller ovigerous female has the carapace 26·5 millim. long.

Wood-Mason established his two species on two specimens, one of which—*L. channeri*—had suffered a good deal from breakage and imperfect re-growth about the frontal region.

A considerable series of the specimens since obtained shows that the two supposed species are really one.

In the Indian Museum collection are numerous specimens, from the Andaman Sea 220 to 271 fms., from the Bay of Bengal 200 to 405 fathoms, from both sides of Ceylon 296 to 406 fms., and from off the Malabar coast, 360 fms.

Uniform salmon-colour in life, white in spirit.

Brachyura vera.

OXYRHYNCHA.

Family *Maiidæ.*

Sub-family *Inachinæ.*

Physachæus, Alcock.

Physachæus, Alcock, Journ. As. Soc. Bengal, Vol. LXIV. pt. 2, 1895, p. 174.

Carapace triangular, inflated (especially in the branchial regions) constricted behind the eyes, ending in a shortish bifid rostrum to the tips of the teeth of which the distal end of the long slender cylindrical basal joint of the antenna of each side is fused.

Eyestalks short, immovably fixed at right angles to the rostrum, eyes well formed but deficient in pigment.

Antennules large, folding longitudinally into the respective hollows of the rostral teeth. Antennæ longer than the carapace and rostrum combined.

Epistome broad, on an oblique plane, separated from the palate by a deep vertical wall. Merus of external maxillipeds obovate, narrower than the ischium, somewhat produced beyond the articulation—near the antero-internal angle—of the palp.

Chelipeds short. Legs slender, very long, ending in filamentous dactyli.

This genus is very closely allied to *Achæus*.

Physachæus ctenurus, Alcock.

Physachæus ctenurus, Alcock, J. A. S. B., Vol. LXIV. pt. 2, 1895, p. 175, pl. iii. figs. 2, 2a-b: Ill. Zool. R. I. M. S. Investigator, Crust. pl. xviii. figs. 1, 1a-b.

Carapace sub-triangular, globosely inflated, with all the regions, except the cardiac, tumid and fairly well delimited, and with a strong post-ocular constriction, beneath which there is an almost vertical descent to the mouth.

The rostrum, which is small, consists of two narrow, slightly divergent, hollow teeth, to either apex of which the distal end of the otherwise perfectly free basal joint of the corresponding antennary peduncle is fused.

Two large erect procurved spines occur in the middle line of the carapace; one on the posterior part of the gastric region, the other behind the cardiac region: on either side of the former, but in a plane anterior to it, there may sometimes be a spinule.

In both sexes the abdomen consists of 5 pieces and is bluntly but strongly carinated down the middle line. The carina in the case of the male ends on the 6th tergum in a huge recurved spine: in the female instead of a spine there is a transverse row of four spinules, and sometimes also a tubercle or a second transverse row of spinules.

The eyestalks are very small, and are rigidly fixed at right angles to the rostrum: the corneæ are almost devoid of pigment. There are no orbits or orbital spines.

The antennæ are distinctly exposed from their base, and are half as long again as the entire carapace, between one-third and two-fifths of their extent being formed by the slender peduncle. The basal joint is slender and almost cylindrical: it is quite free from neighbouring parts, except at the distal end, which is fused with the tip of the rostrum. The flagella are fringed with long hairs.

The antennules are large, and fold longitudinally within the hollow teeth of the rostrum.

The chelipeds in the adult male are considerably more than $1\frac{1}{2}$ times the total length of the carapace and are considerably stouter than the legs, while in the adult female they are considerably less than $1\frac{1}{2}$ times the length of the carapace and are hardly stouter than the legs. Except in respect of the fingers they have much the same form as, though slenderer proportions than, those of *Stenorhynchus*, but the arm is much more strongly and elegantly curved: the arm and wrist are moderately inflated, the former joint, like the ischium, having its lower edge more or less granulate: the palm is compressed, with the edges denticulate: the fingers are strongly compressed, and have the cutting edges accurately and completely apposable throughout, being almost imperceptibly denticulate near the tips only.

In the female the chelipeds have the same general form as in the male, but differ in being much slenderer and in having the lower edge of the ischium and merus strongly spinate.

The legs are slender and filiform, about one-fourth of their length being contributed by the filamentous dactylus: those of the second pair are the longest, being about four times the length of the carapace, rostrum included, and more than two-and-a-half times the length of the chelipeds.

In both sexes the first segment of the sternum is modified to assist in reproduction, being sharply compressed and produced in a vertical direction downwards: in the male, the carina so formed is notched in the middle line so as to form, with the tip of the abdomen, a sort of tunnel for the enormously developed first pair of abdominal legs: in the female it is either notched or entire and forms a high front wall to the brood-pouch, and the openings of the oviducts are pushed forwards so as to lie immediately behind it.

The length of the average adult female carapace is 8·5 millim., the breadth 7 millim. The adult male is a little smaller and has the anterior pair of abdominal legs enormously developed.

Numerous males and egg-laden females have been taken, in the Andaman Sea 185 to 375 fms., and off the Malabar coast, 360 and 406 fms.

The eggs are few in number and are singularly large, being over a millimetre in diameter.

Physachæus tonsor, Alcock.

Physachæus tonsor, Alcock, J. A. S. B., Vol. LXIV. pt. 2, 1895, p. 176, pl. iii. fig. 3; Ill. Zool. R. I. M. S. Investigator, Crust. pl. xviii. figs 2, 2a.

The female, which is the only sex represented in the collection, differs from the female of *Physachæus ctenurus* in the following particulars:—

(1) the gastric region of the carapace, instead of a single large spine, has several smooth tubercles; and the large spine behind the cardiac region is coarser, and is recurved instead of procurved: the post-ocular constriction is less marked:

(2) the abdominal carina ends in a spine, and the sixth tergum has its after edge perfectly smooth instead of quadrispinate:

(3) the eye-stalks are larger, and are compressed instead of cylindrical:

(4) the chelipeds are relatively stouter, being of much the same proportions as those of the male of *Physachæus ctenurus:* the arm is compressed and has its lower border very strongly and sharply carinated: the hands are much thinner and more compressed; the palm having its lower edge, and the fingers their outside edges, sharply cristate:

(5) the first pair of legs, not the second, are the longest, and considerably so.

Length of carapace 11 millim. Breadth of carapace 9·5 millim.

Two egg-laden females from the Andaman Sea, 271 fathoms.

The eggs, as in the preceding species, are large and few in number.

ECHINOPLAX, Miers. *Amended.*

Echinoplax, Miers, Challenger Brachyura. p. 31: Alcock, J. A. S. B., Vol. LXIV. pt. 2, 1895, p. 178.

Carapace piriform, inflated, profusely spiny, somewhat constricted behind the eyes.

Rostrum composed of a pair of spiny horns, between which is a large, vertically deflexed, bifid interantennulary spine.

Eyestalks perfectly retractile against the sides of the carapace, protected, but hardly concealed, by a series of orbital spines: eyes well formed and well pigmented.

The antennules fold nearly longitudinally. The basal antenna-joint is slender, subcylindrical, and of good length; though immovable it is perfectly independent; the antennal flagella are long.

Epistome broad. The external maxillipeds do not completely cover the buccal cavern, but leave the mandibles exposed between them: the merus is slightly narrower than the ischium, and the coarse palp articulates with its antero-internal angle.

Chelipeds in the male long and stout, with enlarged hands: in the female short, and not stouter than the legs, with slender hands.

Legs cylindrical, long, spiny; their dactyli stout and rather short. The last two pairs much shorter than the first two.

All seven abdominal terga are distinct in both sexes, but in the adult male the last three, and in the adult female the last two, move as one piece.

I think that *Echinoplax* is identical with *Ergasticus*, A. M. Edw.

Echinoplax pungens, Wood-Mason.

Echinoplax pungens, Wood-Mason, Ann. Mag. Nat. Hist., March, 1891, p. 259: Alcock, J. A. S. B., Vol. LXIV. pt. 2, 1895, p. 179: Ill. Zool. R. I. M. S. Investigator, Crust. pl. xvii. fig 1. (adult female): and pl. xxxix. (adult male).

Carapace piriform, convex, with the regions well delimited and the branchial and gastric regions inflated; densely covered, as are also the sterna, chelipeds, ambulatory legs, and external maxillipeds, with pungent acicular spines. The abdominal terga of the male and young female are also similarly spiny, but in the adult female they become only distantly and coarsely granular. In the adult male the spines on the last 3 joints of the ambulatory legs are little more than sharp granules.

The rostrum consists of two slender curved divergent spines—less than one-third the length of the carapace proper—the outer and lower surfaces of which are extremely spiny.

The eye-stalks, which have the anterior surface closely spinulate, are retractile, but not to the extent of concealment: there is a strong post-ocular spine—to which, however, the retracted eye does not nearly reach—and numerous smaller spines along the supra-ocular and infra-ocular margins. The antennæ are visible from above, from the middle of the second joint of the peduncle: the peduncle is spiny, with all the joints very slender: the flagellum reaches a little beyond the tip of the rostrum.

The interantennulary spine is large and deeply bifid.

The chelipeds, in the adult male, are twice the length of the combined carapace and rostrum and are much stouter than the legs: nearly half their length is formed by the enlarged somewhat club-shaped hand and the fingers, the fingers being about two-thirds as long as the hand (palm) and being curved so as to meet only at the distal end.

In the female and young male, the chelipeds are about equal in length to the combined carapace and rostrum and are not stouter than the legs.

The legs are cylindrical and end in a stout cylindrical elegantly plumed dactylus, which terminates in a sharp claw. The first pair are the longest, being not far short of three times the length of the carapace and rostrum in the adult male, though much shorter than this in the female. The last two pair of legs are much the shortest in both sexes, being less than twice the length of the carapace and rostrum.

The carapace of an adult male is 114 millim. long and 85 millim. broad: that of an adult female is 79 millim. long and 57 millim. broad.

Dr. A. R. Anderson describes the animal in life, as having its under surfaces chalky white, the upper surface of the carapace pale orange and grey, the upper surface of the legs orange and white, the cornea bluish grey, and the eggs reddish brown.

Our numerous specimens, which represent both sexes at various ages come from numerous stations in the Andaman Sea and from depths varying from 112 to 250 fathoms.

The latest additions are two adult males and an adult female from 185 fms., dredged by Dr. A. R. Anderson, from which an amended diagnosis of the genus has been drawn up.

Echinoplax rubida, Alcock.

Echinoplax rubida, Alcock, J. A. S. B., Vol. LXIV, pt. 2, 1895, p. 179: Ill. Zool. Investigator, Crust. pl. xvii, figs. 2, 2*a*.

Differs from the preceding species in the following characters, specimens of the same sex and of approximately the same size being compared:—

(1.) The branchial regions are not inflated above the level of the cardiac region, and the latter region is therefore larger and better defined:

(2.) The carapace, instead of being everywhere covered with pungent acicular spines of uniform size, is finely granular, with certain definitely placed distant thornlike spines of conspicuous magnitude, namely:—four in triangle on the gastric region, two side by side on the cardiac region, two side by side on the intestinal region, three on each hepatic region, and three on each epibranchial region: besides these there are some smaller spines on the lateral aspect of the pterygostomian and branchial regions, and a very large spine at either antero-external angle of the buccal cavern:

(3.) The rostral spines are less divergent, and have elegantly curved tips:

(4.) The abdominal terga (of the female), instead of being everywhere closely covered with pungent spines, are merely finely and distantly granular, with a single large spine on the first tergum, and a pair of smaller spines on the second, in the middle line:

(5.) The legs are much less spiny, the propodites of the ambulatory legs being fringed with stiff bristles instead of spines:

(6.) The colour differs, being, in spirit specimens, a warm brown, instead of a pale yellow.

A single female specimen, apparently (from the condition of the brood-pouch and the form of the abdomen) an adult, from the Andaman Sea, 90 to 177 fms.

The length of the carapace is 35 millim., its breadth 21 millim.

Cyrtomaia, Miers.

Cyrtomaia, Miers, Challenger Brachyura, p. 14: Mary J. Rathbun, Proc. U. S. Nat. Mus. XVI. 1893 (1894), p. 228: Alcock, J. A. S. B. Vol. LXIV. pt. 2, 1895, p. 163.

Very closely related to *Echinoplax*, from which it appears to differ only in the following respect:—

(1) The carapace instead of being profusely spiny has a few large definitely placed spines, hence the orbital margin is not so spiny.

(2) The vertically deflexed interantennulary spine is simple.

(3) The rostral spines are shorter.

I very much doubt whether these characters are of generic value; and I think that if adults of all three forms could be compared, both *Cyrtomaia* and *Echinoplax* would be placed under A. Milne Edwards' genus *Ergasticus*.

Cyrtomaia suhmi, Miers, *var.*

Cyrtomaia suhmi, Miers, Challenger Brachyura, p. 16, pl. iii. fig. 3: Mary J. Rathbun, P. U. S. Nat. Mus. XVI. 1893 (1894), p. 230.

This species was described from a damaged male dredged by the Challenger in 500 fms., between the Philippines and Moluccas. Our specimen, which is a male in good enough preservation, was dredged in 430 fms., off the Travancore coast. It differs from the description of the type in the following characters:—

(1) There is a third spine, much smaller than the other two, and standing in the middle line, on the gastric region.

(2) On the cardiac region, instead of a single spine, there are two spines standing on a common tubercle.

(3) The last abdominal tergum is quite smooth.

These slight differences hardly warrant a specific designation.

PLATYMAIA, Miers.

Platymaia, Miers, 'Challenger' Brachyura, p. 12. ♀: Alcock, J. A. S. B., Vol. LXIV. pt. 2, 1895, p. 180 ♂ & ♀.

Carapace sub-orbicular. Rostrum short, tridentate owing to the size and projection of the interantennulary septum. No pre-ocular spine; but a post-ocular spine against which the eye is retractile, but which affords no concealment to the eye. Epistome extremely narrow. Eyes large, with short eye-stalks. Basal antenna-joint short, cylindrical, and perfectly free: the flagellum and part of the peduncle visible from above.

External maxillipeds with the meropodite narrow, and bearing the next joint at its summit. Chelipeds in the male long, with a long inflated club-shaped palm: in the female very short and slender. Ambulatory legs long, with remarkably thin compressed joints: some of the legs spiny.

Abdomen in both sexes with all the segments separate.

This genus appears to be very closely related to *Macrochira*.

Platymaia wyville-thomsoni, Miers.

Platymaia wyville-thomsoni, Miers, 'Challenger' Brachyura, p. 13, pl. ii. fig. 1.

Platymaia wyville-thomsoni, Wood-Mason and Alcock, Ann. Mag. Nat. Hist., March, 1891, p. 258, and May. 1891, p. 401: Alcock, J. A. S. B., Vol. LXIV. pt. 2, 1895, p. 181: Illustrations of the Zoology of the Investigator, Crust. pl. xvi. (adult male natural size).

Carapace transversely sub-circular with the cervical grove well defined: its surface ranging from spinate (in the young) to nearly smooth (in old adults). The rostrum, which is so short as not to break beyond the general outline, consists of three stout spines of equal size, the middle one being the horizontally projecting interantennulary spine.

The hepatic region of the carapace bears (in the adult) a nearly vertically disposed row of three spines, against the upper one of which the eye is retractile.

The eyestalks are short, and the eyes large and oval. The antennæ are about one-third the length of the carapace, and are plainly visible, in almost the whole of their extent, from above: the joints of the peduncle are short slender and cylindrical, the basal joint being perfectly free.

The external maxillipeds have the meropodite narrow (about half the breadth of the ischiopodite) and giving attachment to the coarse palp at the summit: both meropodite and ischiopodite are spiny.

The chelipeds vary considerably according to sex: in both sexes they are spiny up to the base of the fingers; but whereas in the female and young male they are much slenderer than any of the legs and are not longer than the carapace, in the adult male they are from two to three times the length of the carapace and are much stouter than any of the legs—especially as regards the palm, which is swollen and club-shaped.

The legs are long and slender, with the joints thin and compressed, the propodites being blade-like. The 1st pair, which are from $3\frac{3}{4}$ (female) to $5\frac{1}{2}$ (male) times the length of the carapace, are remarkable for their propodite and dactylus, the front edge of which bears a double comb of enormous spines, the posterior edge also being spinulate: both edges of the merus and carpus also are distantly spinulate. The 2nd and 3rd pairs have the front edge of the merus distantly spinulate, and they, as well as the 4th pair, have the front edge of the razor-like merus closely fringed with long stiff hairs.

The abdomen in both sexes is seven-jointed, the abdominal terga, like the thoracic sterna, bearing a few spines or tubercles. The epimeral plates corresponding to the third and fourth trunk legs are also spinate.

Andaman Sea, 130—405 fathoms.

In very young specimens (carapace less than half an inch in diameter) the whole carapace is closely and sharply spiny.

In larger specimens (carapace about three-quarters of an inch in diameter) the carapace has become closely and finely granular, with the spines persistent only in definite situations, somewhat as in Miers' figure and description, (*loc. cit.*).

In larger specimens (carapace two and a half inches in diameter) the carapace has become coarsely and bluntly granular, without any spines, except a few quite anteriorly in the neighbourhood of the hepatic region.

In the largest specimens (carapace three to nearly four inches in diameter) the carapace is in places quite smooth, the only spines present being two external to the eye, and one on the front margin of the hepatic region.

In contrast with the carapace, the spines on the abdominal sterna of the male show no signs of effacement with age.

The colours also seem to vary with age. In young males taken by myself the carapace was red, with or without red points, and the legs were red and white in alternate bands. In adults of both sexes taken by Dr. A. R. S. Anderson the upper surface of the carapace was bright orange, the legs were banded alternately dark and pale orange, and the under surface was bluish.

ENCEPHALOIDES, Wood-Mason.

Encephaloides, Alcock, J. A. S. B. Vol. LXIV. pt. 2, 1895, p. 186.

Carapace, owing to the remarkable inflation of the branchial regions, heart-shaped and posteriorly as broad as long (rostrum included): the branchial regions rising up and meeting across the carapace in the middle line. Rostrum simple, shaped like the beak of a bird. Eyes retractile against the sides of the carapace: a small pre-ocular and post-ocular spine, but no definite orbit.

Basal antenna-joint slender throughout but flat and fused with the rostrum: the antennæ visible, dorsally, from the base of the second joint.

The merus of the external maxillipeds is produced antero-externally to form a foliaceous lobe which covers the greatly produced efferent branchial orifice; it is thus both longer and much broader than the ischium.

Abdomen in the male seven-jointed: in the female the fourth, fifth and sixth segments, though distinctly recognizable, are firmly fused together.

Chelipeds in both sexes slender. Legs long and slender.

Only eight branchiæ on each side.

Encephaloides Armstrongi, Wood-Mason.

Encephaloides armstrongi, Wood-Mason, Ann. Mag. Nat. Hist. March, 1891, p. 259: Alcock, J. A. S. B., Vol. LXIV. pt. 2, 1895, p. 187: Ill. Zool. Investigator, Crust, pl. xix. figs. 2, 2*a*.

Carapace heart-shaped: its greatest breadth is equal to its length with the rostrum: its surface in the adult is nodular or pustular, in the young coarsely spiny. The gastric and hepatic regions are well-defined: but the cardiac and intestinal regions are entirely concealed by the branchial regions, which rise up like a pair of mammæ, and meet, but without any fusion of walls, down the middle line.

The rostrum, which is shaped exactly like the beak of a bird, is about one-fourth the length of the carapace proper, measured from the pre-ocular spine, and has a finely serrated edge.

In the male the abdomen is distinctly seven-jointed; but in the female the fourth, fifth and sixth segments are immovably sutured together.

The eyes which are small, slender, and ill-pigmented, are retractile against the side of the carapace: there is a very narrow supra-orbital eave ending anteriorly in a minute tooth, and there is a small post-ocular spinule.

On the dorsal aspect the antennæ are plainly visible on either side of the rostrum, from the base of the 2nd joint of the peduncle: the flagella, which are of hairlike tenuity, hardly surpass the tip of the rostrum.

Owing to the prolongation of the efferent branchial canals, the front edge of the buccal cavern is V-shaped, the merus of the external maxillipeds ear-shaped, and the antero-external angles of the buccal cavern are puffed out beyond the level of the retracted eye.

The legs and chelipeds are slender, cylindrical, and smooth; and are so much longer in the adult male than in the adult female as to need a separate description for each sex.

In the *male* the chelipeds are just over twice the length of the carapace and rostrum; the first pair of legs, which are about half a dactylus length longer than the second, are four and a half or nearly five times the length of the carapace and rostrum, or about two and a quarter times the length of the chelipeds, the third pair of legs are between two and three quarters and three times the length of the carapace and rostrum, and the fourth pair are only about one and two-thirds times the length of the carapace and rostrum: the fingers are not quite a third the length of the palm, which is a slender cylindrical joint hardly broader than the arm.

In the *female* there is no such great disproportion between the legs: the chelipeds are only about one and a fifth times the length of the carapace and rostrum; the first pair of legs are only about twice and a quarter, the second pair only about twice, the third pair about once and two-thirds, and the fourth pair about once and a quarter the length of the carapace and rostrum: the fingers are not quite half the length of the palm.

The carapace in the adult male is 42 millim. long and 42 millim. broad, in the adult female 33 millim. long and 32 millim. broad.

The colours in the fresh state are, carapace pinkish yellow, legs pink.

This species is characteristic of depths between 60 and 100 fathoms in the Bay of Bengal, and there are scores of specimens in the Indian Museum.

Encephaloides Rivers-Andersoni, n. sp.

Differs from the preceding species in the following particulars:—

(1) The carapace is more elongate and narrower, on account not only of the greater relative length of the rostrum, but also of the less puffed out pterygostomian regions.

(2) Although the angles of the buccal cavern are strongly produced, they do not reach beyond the level of the retracted eye.

(3) The eyes are much paler.

(4) In the (?adult) male, though not in the adult female, the rostrum is much longer, its length, measured from the pre-ocular spine, being nearly half that of the rest of the carapace.

(5) In the (? adult) male also, though not in the female, the palm instead of being cylindrical is compressed and appreciably broadened so as to be conspicuous as the stoutest joint in the whole series of legs.

Length of carapace of male 26 millim., breadth 21 millim. Length of carapace of female 22 millim., breadth 18 millim.

Ten specimens from off Travancore coast, 406 fms.

This species closely resembles *E. armstrongi*, but the shape of the carapace, especially of the pterygostomian regions, is quite different in both sexes; and in the male the form of the rostrum and of the hands are also quite different.

Subfamily *Pisinæ*.

Sphenocarcinus, A. Milne Edwards.

Sphenocarcinus, A. Milne Edwards, Miss. Sci. Mex., Crust., I., p. 135: Miers, Journ. Linn. Soc., Zool., Vol. XIV. 1879, p. 663, and 'Challenger' Brachyura, p. 34: Alcock, J. A. S. B., Vol. LXIV. pt. 2, 1895, p. 193.

Carapace elongate sub-pentagonal, broad behind, tapering in front to a huge rostrum formed of two spines (fused together to near the tip). The surface of the carapace is symmetrically and deeply honey-combed by broad deep channels which leave symmetrical tubercles with over-hanging edges between them.

There are no true pre-ocular and post-ocular spines, but the eye is deeply sunk between two low smooth excrescences which are pre-ocular and post-ocular in position.

The basal antenna-joint is truncate-triangular, and the antennary flagella are completely hidden beneath the rostrum.

The epistome is long and narrow. The external maxillipeds have the merus as broad as the ischium, somewhat dilated at the antero-external angle, and somewhat excavated at the antero-internal angle for the insertion of the small palp.

The chelipeds are not much stouter, and not much shorter than the next pair of legs, which are the longest: the dactyli of the legs, though stout recurved and prehensile, are not toothed along the posterior edge.

Abdomen, in both sexes, seven-jointed.

Sphenocarcinus cuneus (Wood-Mason).

Oxypleurodon cuneus, Wood-Mason, Ann. Mag. Nat. Hist., (6) VII. 1891, p. 261.

Sphenocarcinus cuneus, Alcock, J. A. S. B., Vol. LXIV. pt. 2, 1895, p. 193: Ill. Zool. Investigator, Crust. pl. xxi. figs. 1, 1a.

Carapace elongate sub-pentagonal, narrowing to a long tapering cylindrical rostrum, which, in the male, is longer than the carapace and only emarginate at the extreme tip, but, in the female, is shorter than the carapace and distinctly bifid at the end.

The carapace is symmetrically honey-combed by deep channels, which leave between them great symmetrically undermined islets, as follows:—one, very elongate-oval, on the gastric region; one, triangular, on the cardiac region; one, somewhat semilunar with one horn much produced laterally, on each branchial region; and one, Cupid's bow-shaped, along the posterior border. Besides these there are some smaller islet-like excrescences, namely, on each side, a supra-ocular, post-ocular, hepatic, and branchial.

Between the supra and post-ocular excrescences, are set the small squat little-movable eyes.

Of the trunk-legs, the 2nd pair (*i.e.*, first ambulatory legs) are the longest, being very slightly longer than the chelipeds, and considerably shorter than the carapace measured with the rostrum, but much longer than any of the last 3 pairs of legs.

In the female all the long joints, except the dactyli, and in the male all except the dactyli and propodites, both of chelipeds and of legs, are strongly carinated dorsally: in the case of the carpopodites there is a second dorsal carina.

The chelipeds are hardly stouter than the next pair of legs, except as regards the palm in the male, which is broadened and somewhat inflated. In neither sex are the short white polished fingers apposable throughout.

		Male.		Female.
Length of carapace and rostrum	...	19 millim.	...	18·5 millim.
Greatest breadth of carapace	...	12 „	...	13 „
Length of rostrum alone	...	10·5 „	...	8·7 „
Length of 2nd pair of trunk-legs	...	15·5 „	...	15 „

From the Andaman Sea, 161 to 250 fathoms.

OXYPLEURODON, Miers.

Oxypleurodon, Miers, Challenger Brachyura, p. 38.

The only difference that I can recognize between *Oxypleurodon* and *Sphenocarcinus* is that in *Oxypleurodon* the rostral spines are divergent instead of being approximated. Such a slight difference cannot be of generic value.

Sphenocarcinus (Oxypleurodon) stimpsoni, Miers.

Oxypleurodon stimpsoni, Miers, Challenger Brachyura, p. 38, pl. vi. fig. 1: Anderson, J. A. S. B., Vol. LXV. pt. 2, 1896, p. 106.

A small male apparently of this species was dredged off Colombo in 180–217 fms.

Dr. Anderson records the colour of the living animal as orange.

SCYRAMATHIA, A. Milne Edwards.

Scyramathia, A. Milne Edwards, Compt. Rend. XCI. 1881, p. 356: G. O. Sars, Norwegian North-Atlantic Expedn., Crust. I⁴. p. 5: S. I. Smith, Albatross Crustacea (1884) 1886, p. 21: Miers, Challenger Brachyura, p. 25: A Milne Edwards and Bouvier, Hirondelle Brachyures et Anomures (Monaco, 1894) p. 12: Faxon, Albatross Stalk-eyed Crust., Mem. Mus. Comp. Zool. Harvard, Vol. XVIII. 1895, p. 10: Alcock, J. A. S. B. Vol. LXIV. pt. 2, 1895, p. 201.

Carapace piriform or elongate-triangular, armed either with tubercles, or with long spines much like those of *Anamathia* in their uniform size and definite arrangement: the hepatic and lateral epibranchial spines are always prominent and very conspicuous. The rostrum consists of two spines, which are usually long and slender. The eyes are small, and are retractile against a sharp post-ocular process which commonly is but little cupped: there is also a supra-ocular eave which terminates either in a forwardly directed tooth or in an upturned spine. Basal antenna-joint not very broad, sharply truncated: the mobile portion of the antennæ freely exposed on either side of the rostrum.

Merus of the external maxillipeds as broad as the ischium, slightly expanded at the antero-external angle, and bearing the palp at the antero-internal angle.

Chelipeds in the adult male (but not in the female and young male) enlarged, with the palms broadened and compressed.

First pair of ambulatory legs markedly the longest.

The abdomen in both sexes consists of seven distinct segments.

The chief difference between *Scyramathia* and *Anamathia* is that in the former the supra-orbital eave is broader and is prolonged forwards as a tooth or spine. In the adult male of *Scyramathia* also, though not in the female and young male, the hands are compressed and somewhat enlarged, whereas in *Anamathia rissoana* (of which there is a male specimen in the Indian Museum) the hands are slender and almost cylindrical.

Whatever may be the value of these differences, our species are all, I think, congeneric with the *Amathia carpenteri* of Norman and with the *Anamathia pulchra* of Miers.

I am inclined to think that the *Pugettia velutina* described and figured by Miers in the Challenger Report, also belongs to this genus.

Scyramathia pulchra, Miers.

Anamathia pulchra, Miers, 'Challenger' Brachyura, p. 26, pl. iv, fig. 1 (adult male).

Anamathia livermorii, Wood-Mason, Ann. Mag. Nat. Hist. March 1891, p. 260 (young male and adult female)

Scyramathia pulchra, Alcock, J. A. S. B., Vol. LXIV, pt. 2, 1895, p. 202: Ill. Zool. Investigator, Crust. pl. xiv fig. 3 (female).

Body and limbs everywhere closely covered with short hairs, which on the carapace are peg-shaped; and with numerous long scattered setæ. The carapace, which is subpiriform, is armed with twenty long sharp spines disposed in five longitudinal series. Of these spines six are on the gastric region, one is on the cardiac, and one on the intestinal region, one stands above either eye, one on each hepatic, and four on each branchial region: in addition there is a distinctly cupped post-ocular lobe.

The rostrum consists of two slender divergent spines, the length of which is more than half that of the carapace proper.

The eyes are small, and the cornea, though retractile against the post-ocular lobe, can never be concealed.

The basal antennal joint is broad, and has its antero-external angle somewhat produced: the mobile portion of the antenna is completely exposed to dorsal view.

The external maxillipeds have the ischium and merus somewhat concave.

The chelipeds vary according to sex. In the adult male they are longer than the carapace and rostrum, and are far stouter than any of the legs: the wrist is enlarged and sculptured with carinæ, the palm is broadened, as well as somewhat carinate along both edges and strongly produced at the postero-inferior angle, and the fingers are apposable in their distal half only. In the female and young male they are shorter than the carapace with the rostrum, and are hardly stouter than the other legs; all the joints are subcylindrical, and the fingers are apposable in the greater part of their extent.

In both sexes, the merus of all the legs, including the chelipeds, has a spine or tooth at the far end of its upper margin. The first pair of true legs, which are the longest, are, in the male, nearly twice the length of the carapace and rostrum, but in the female are considerably shorter.

From the Andaman Sea, 130 to 561 fathoms.

Numerous specimens of both sexes are in the Indian Museum.

The eggs are very large.

Scyramathia Rivers-Andersoni, Alcock.

Scyramathia Rivers-Andersoni, Alcock, J. A. S. B., Vol. LXIV. pt. 2, 1895, p. 203: Ill. Zool. Investigator, Crust. pl. xxii. figs. 2, 4, 4a.

Carapace closely covered with peg-shaped hairs with long setæ interspersed: legs with few setæ. The carapace, which is piriform and somewhat inflated, has, besides a supra-ocular tooth and a sharp post-ocular process, and besides a salient hepatic spine, and a still more salient lateral epibranchial spine (about two-fifths the greatest breadth of the carapace in length), six sharply conical tubercles evenly and equidistantly arranged in a circle round a central cardiac tubercle: of these the most posterior overhangs the middle of the posterior border, while the most anterior, which is situated far back on the gastric region, is flanked on either side by a very faint eminence.

The rostrum consists of two slender divergent horns, the length of which in the male is about three-quarters, in the female about two-thirds, that of the rest of the carapace.

The eyes are small and pale, and though freely movable forwards are not retractile backwards further than to impinge against the summit of the post-ocular process of the carapace. The basal antennal joint, which is of no great width, is sharply truncated: the mobile portion of the antenna is freely exposed on either side of the rostrum.

The chelipeds in the fully adult male (but not in the young male) are much stouter than the legs, and are as long as the carapace and rostrum; the arm is prismatic with knife-like edges, the upper edge ending in a spine; the wrist is bicarinate, the outer carina being very prominent; the hands, which form nearly half their total length, have the palm carinate along the upper edge, and the fingers slightly separated when closed.

In the female the chelipeds are not stouter than the legs, are not much longer than the carapace proper, and have the fingers closely apposable throughout.

Of the ambulatory legs the first are much the longest, being nearly half again as long as the carapace and rostrum; while the last two pairs are very short and have their dactyli reduced in length, increased in strength, and strongly recurved.

	Male.	Female.
Length of carapace and rostrum	21 millim.	16·5 millim.
" rostrum ...	9 "	7 "
" chelipeds	21 "	11 "
" 2nd pair of trunk-legs	31 "	20 "
" 5th ...	15 "	11 "

Off Malabar coast, 406 and 430 fms.

The next two species more resemble the *Pugettia velutina*, Miers (Challenger Brachyura, p. 41, pl. vi. fig. 2).

Scyramathia beauchampi (Alcock and Anderson).

Anamathia beauchampi, Alcock and Anderson, J. A. S. B., 1894, pt. 2, p. 185.

Scyramathia beauchampi, Alcock, J. A. S. B. Vol. LXIV. pt. 2, 1895, p. 204, Ill. Zool. Investigator Crust. pl. xx. fig. 2, 2a.

Body and legs downy, and with numerous large coarse curly clavate hairs, which are very regularly arranged on the legs, where also they are coarsest and closest. Carapace sub-piriform, with the following armature:—

On either hepatic region a great up-curved earlike spine (without any bullous base). On either branchial region, posteriorly, a strong up-turned spine; and anteriorly, near the middle line, a smaller coarse tooth. On the gastric region four sharpish tubercles. On the narrow sunken cardiac region a coarse sharp tooth. On the posterior border, in the middle line, a coarse granule.

The rostrum consists of two more (♀) or less (♂) divergent spines, the length of which is about one-third that of the rest of the carapace.

The eyes are small, and are almost devoid of pigment: they are to some extent hidden beneath a pre-ocular tooth of moderate dimensions, and are retractile against a larger laterally-compressed foliaceous post-ocular lobe.

The antennæ are completely exposed from the base of the second joint of the peduncle.

The chelipeds in the male are massive, and in length are more than half again as long as the carapace and rostrum: all their joints, from the ischium to the propodite, have one or more of their edges conspicuously and sharply cristiform, this being specially well marked in the case of the long trigonal arm, which has all its edges sharply phalanged, and in the case of the equally long slightly inflated palm, which has razor-like edges. The fingers which are not nearly half the length of the palm, are acute, and have their cutting edges entire.

The legs are slender, with cylindrical joints, the first are nearly or quite equal in length to the chelipeds, the others decrease gradually in size.

In an adult female, equal in size to the male above described, the chelipeds are shorter than the 2nd pair of legs, and are similar in general proportions to the other legs.

Colours in life: "Earth-colour with the chelipeds pink."

	Male.	Female (adult.)
Length of carapace (including rostrum)...	18 millim. ...	15·5 millim.
Greatest breadth of carapace ...	12·5 „ ...	11·5 „
Length of cheliped ...	29 „	14 „
Greatest breadth of palm	4·5 „	1 „

From the Bay of Bengal, 193 and 210 fathoms.

The ova are large (diam. 1 millim.) and rather few in number.

In young males the chelipeds are of proportions intermediate between those of the adult male and female.

Scyramathia globulifera, Wood-Mason.

Pugettia globulifera, Wood-Mason, Ann. Mag. Nat. Hist. March, 1891, p. 260.

Scyramathia globulifera, Alcock, J. A. S. B. Vol. LXIV. pt. 2, 1895, p. 205: Ill. Zool. Investigator, Crust. pl. xx. figs. 3, 3a.

Distinguished by the vertically erect ear-like hepatic spine, the base of which forms a great polished bulla on either side of the buccal frame, giving the animal, when viewed front end on, a bat-like appearance.

The body and legs are downy, the legs being fringed with short broad curly hairs.

The carapace, in which the cardiac region is broad and prominent and not, as in *S. beauchampi*, narrow and sunken, has, besides the hepatic spine already mentioned, the following marks :—

On the branchial regions, below and anteriorly, a sharp sinuous human-ear-shaped crest; above and posteriorly a spine; and near the middle line, anteriorly, an acumination. On the gastric region four faint elevations. On the cardiac region, and also on the intestinal region, in the middle line, an acuminate eminence.

The rostrum consists of two divergent spines, about one-third the length of the rest of the carapace.

The eyes stand well out from beneath the pre-ocular spine, and are retractile against a small post-ocular tooth.

The other appendages closely resemble those of the preceding species; but the chelipeds, in the adult male, are shorter, being only equal in length to the carapace and rostrum, and the fingers have their cutting edges crenulate instead of smooth.

In females and in young males the chelipeds have the same relative proportions as in *Scyramathia beauchampi*.

	Male.	Female (adult).
Length of carapace (including rostrum) ...	17 millim. ...	13 millim.
Greatest breadth of carapace ...	10 „ ...	7·5 „
Length of cheliped	18 „ ...	9·5 „
Greatest breadth of palm	4 „	1·2 „

From the Andaman Sea, 130–240 fathoms.

Sub-family *Maiinæ*.

Maia (Lamk.) Edw.

Maia, Lamarck, Syst. Anim. sans vertéb. V. 240 (*partim*): Bosc, Hist. Nat. Crust. I. 245 (*partim*): Latreille, Hist. Nat. Crust. VI. 87 (*partim*): Leach, Malac. Pod. Brit. pl. xviii and text: Desmarest, Consid. Gen. Crust.,

p. 143: Risso, Hist. Nat. Eur. Merid. Vol. V. p. 22: Latreille, in Cuvier Règne An. ed. 2, 1829, p. 59: Milne Edwards, Hist. Nat. Crust., I. 325: Dana, U. S. Expl. Exp. Crust. pt. I. p. 78: Bell, British Stalk-eyed Crust. p. 39: Miers, Journ. Linn. Soc., Zool., Vol. XIV. 1879, p. 655: Alcock, J. A. S. B., Vol. LXIV. pt. 2, 1895, p. 238.

Carapace piriform, with the regions indistinct, the surface closely granular or spinular, and the lateral borders usually armed with large spines. The rostrum consists of two rather short, straight, divergent spines. The basal joint of the antennæ is broad, and has both the antero-external and antero-internal angle produced to form spines: the mobile portion of the antenna, which appears to spring from within the orbit, is completely exposed. The eye-stalks are long and curved, and bear the cornea chiefly on their ventral surface. The orbit is formed by a prominent supra-ocular eave which has its postero-external angle produced, by a sharp post-ocular spine, and by another spine between these two: the eyes are completely concealed from dorsal view when retracted. The external maxillipeds have the merus as broad as the ischium, the palp being attached to the antero-internal angle of the merus.

The chelipeds are slender, with cylindrical joints and styliform fingers. The ambulatory legs decrease very gradually in length: the first pair are not much longer than the carapace and rostrum: the dactyli of all are styliform.

The abdomen in both sexes consists of seven distinct segments.

Maia gibba, Alcock.

Maia gibba, Alcock, J. A. S. B. Vol. LXIV. pt. 2, 1895, p. 239, pl. iv. fig. 5: Ill. Zool. Investigator, Crust. pl. xxi. figs 5, 5a.

Very near *Maia miersii*, Walker (J. L. S., Zool., Vol. XX. 1890, p. 113, pl. vi. figs. 1-3).

Distinguished (1) by the globose inflation of the posterior (branchiostegal) part of the closely and crisply tubercular carapace, and by the corresponding declivity of the anterior part, giving the animal a hunch-backed appearance; (2) by the absence of *large* marginal spines on the carapace.

Carapace remarkably swollen in its posterior part, where its greatest breadth is from about three-fourths (♂) to seven-eighths (♀) its extreme length with the rostrum; and closely covered with sharp piliferous tubercles, which, in the male, but hardly in the female, become spinular in the middle line and along the lateral borders.

The rostrum, which, like the anterior part of the carapace, is somewhat declivous, ends in two acute divergent hairy spines, which in the male are about one sixth, in the female about one-eighth, the rest of the carapace in length. The eyes and orbits are just as in *M. squinado* (with specimens of which this species has been compared), only the cornea is relatively very much larger and almost entirely ventral in the present species, and the spine between the spine of the pre-orbital-hood and the post-orbital spine is nearly as large as either of these.

The antennæ are in all respects as in *M. squinado*, except that the basal joint is slightly narrower.

The appendages are just as in *M. squinado*—the legs being short and hairy and the chelipeds smooth and polished—with the single difference that the chelipeds are only as long as, and are much slenderer than the *last* pair of legs, and are therefore very much shorter than the first pair, which hardly exceed the carapace and rostrum in length.

	Male.	Female.
Length of carapace ...	32 millim.	41 millim.
Greatest breadth of carapace ...	25 ,,	35 ,,
Length of chelipeds	24 ,,	31 ,,
,, ,, 1st pair of legs	33·5 ,,	46 ,,

From the Andaman Sea, 250 fms.

CORYSTOIDEA.

TRICHOPELTARIUM, A. M. Edw.

Trichopeltarium, A. Milne Edwards, Bull. Mus. Comp. Zool., Vol. VIII. 1880, p. 19.

Said to differ from *Hypopeltarium* in having the carapace strongly convex and hairy and the chelipeds, in the male, very greatly unequal.

? *Trichopeltarium ovale*, Anderson.

? *Trichopeltarium ovale*, Anderson, J. A. S. B. Vol. LXV. pt. 2, 1896, p. 103: Ill. Zool. Investigator, Crust. pl. xxv. figs. 4, 4a.

Carapace egg-shaped, covered with spines which on its dorsal surface are bifid or multifid, and with short stiff not very conspicuous hairs. The regions are well defined by coarse grooves: the gastric is divided into three sub-regions, and the cardiac into two, and on either side of the cardiac region a semilunar area is marked off on the branchial region.

The front, which is cut into three prongs, is about one-seventh the greatest breadth of the carapace, and is separated from the orbit by a deep notch.

The orbits are very incomplete: they are formed by a prominent præocular tooth (parallel with, but less prominent than, the front), below which, at the inner suborbital angle is an almost equally prominent coarse spine: there are also two teeth—one at the external orbital angle, and the other between this and the præocular tooth—hardly distinguishable from the ordinary spines of the carapace. The eyestalks which are slender, tapering, and of good length, do not nearly fill the shallow orbital cavity.

The antennules fold longitudinally in fossæ, beneath the front: their basal joint is large. The antennæ arise almost in the same transverse line with the antennules: their basal joint forms a large part of the floor of the orbit.

The epistome is arched much above the plane of the external maxillipeds. The efferent branchial channels are defined each by an incomplete ridge, and are patulous. The external maxillipeds are slender, and leave the mandibles exposed between them: the merus is obovate and narrower than the ischium, the palp is coarse.

The chelipeds and legs are spiny and bristly, the spines in the case of the legs being well pronounced only on the dorsal surfaces of the meropodites.

In the female the chelipeds are shorter and not much stouter than the legs and are about as long as the carapace.

The legs are little unequal in length, the first pair, which are slightly the longest, being hardly half as long again as the carapace: they all end in long, stout, cylindrical, sharply styliform dactyli.

The abdomen of the female is seven-jointed and covered with coarse hairs: the first two segments are narrow and bear some spines in the middle line, the other segments are broader and on them the spines decrease in size to the seventh segment which is smooth.

The colour in life is recorded by Dr. A. R. Anderson as pale bluish yellow.

Length of carapace and rostrum 64 millim., breadth 55·5 millim, depth 35 millim.

A single female from off the west coast of Ceylon 180–217 fms., on a foul bottom of broken coral.

Trachycarcinus, Faxon.

Trachycarcinus, Faxon, Bull. Mus. Comp. Zool. XXIV. 1893, p. 156: Mem. Mus. Comp. Zool. XVIII. 1895, p. 25.

"Carapace pentagonal, moderately convex, lateral margins long, nearly straight, toothed. Front narrow, produced, three-toothed. Orbits large with forward aspect, imperfect, with two hiatuses above, one below, and one at the inner angle; lower wall formed chiefly by the carapace. Anterior margin of buccal cavity not distinctly defined, epistome short, ridges of the endostome developed. Sternum long and rather narrow. Abdomen of male narrow and five-jointed, the third, fourth, and fifth segments consolidated. Eye-stalks very small, retractile within the orbits. Antennules longitudinally folded. The antennæ lie in the inner hiatus of the orbit; their basal segment is but slightly enlarged, not filling the hiatus at the inner angle of the orbit nor attaining to the front, subcylindrical, unarmed, imperfectly fused with the carapace; the second segment is longer and slenderer than the first, the third segment about equal to the second in length, but slenderer; all these segments are furnished with long and coarse setæ; the whole antenna is less than one half as long as the carapace. The ischium of the outer maxillipeds is produced at its antero-internal angle; the merus of the same appendages is rounded at the antero-external angle, obliquely truncated but not emarginated at the antero-internal angle, where it articulates with the following segment. Legs of moderate length. Right and left chelipeds very unequally developed in the male. Dactyli of ambulatory legs styliform, straight, slender, longer than the penultimate segments."

Trachycarcinus glaucus, Alcock and Anderson, Plate II. fig. 2.

Carapace irregularly pentagonal, its surface coated with short stiff club-shaped hairs; the regions well defined, rather tumid, much subdivided into tumid lobules, of which the convexities are capped by clusters of large conical granules and the general surface also is studded especially in the young with similar granules.

Front narrow, horizontal, prominent, deeply cleft into three prongs of nearly equal size.

Antero-lateral borders half as long again as the postero-lateral, armed with three stout pinnulate spines not including the outer orbital angle: postero-lateral borders entire, posterior border finely beaded.

Upper orbital wall deeply cleft into three pinnulate teeth, lower orbital border deeply concave, its inner angle strongly spiniform. Eyestalks slender, rather long: the eyes, which are more ventral than terminal, are dull and faintly pigmented (as in many species of *Munidopsis*), and are non-facetted.

Antennal flagella short, extremely slender, not hairy.

Chelipeds remarkably unequal in the male, equal in the female.

The smaller cheliped of the male and both chelipeds of the female are about as long as the carapace, and are coated, almost to the finger-tips, with stiff club-shaped hairs, which are short except along the upper border of the wrist and hand and of the basal part of the finger, where they are long: beneath the hairs are some scattered granules, and along the upper border of the arm, wrist and hand are some denticles: the inner angle of the wrist is strongly spiniform, and the far end of the upper border of the hand is dentiform.

The larger cheliped of the male is about twice the length of the carapace, about half its length being formed by the hand and fingers: the greatest breadth of the hand is about half the length of the carapace. It is almost smooth, the upper border of the arm and hand, and the inner border and upper and outer surfaces of the wrist, alone being furnished with denticles and hairs: the inner angle of the wrist is spiniform.

The legs are covered with short stiff club-shaped hairs which are rather more thick-set on the anterior borders and on the dactyli than elsewhere. The second and third pair, which are rather longer than the first and last pair, are somewhat less than $1\frac{3}{5}$ times the length of the carapace. All the dactyli end in a little claw.

The abdomen of the male consists of seven distinct segments, but the 3rd, 4th and 5th move together.

In life the animal is covered with a coat of mud held together by the hairs above described, the only bare parts being the hand and fingers and part of the arm of the larger cheliped of the male.

The colours in life are described by Dr. A. R. Anderson as "white with a bluish tinge, eyes with a slight reddish opalescence." In spirit the bluish tinge is fainter, the eyes are a pale milky yellow-ochre, and the large hand is ivory-white.

The dimensions of the largest male are as follows :—

Length of carapace	18·5	millim.
Breadth of carapace	14·5	"
Combined length of hand and fingers, along lower border	14·75	"
Combined length of basal joints arm and wrist, along upper border	15	"

Fifteen specimens were dredged off the Travancore coast at a depth of 430 fms. The bottom consisted chiefly of coral (living and dead).

Several of the specimens were egg-laden females. The eggs are comparatively few in number and are large, their diameter being about 1·3 millim.

This species is very like *Trachycarcinus corallinus*, Faxon, which was dredged by the "Albatross" off Panama and the Pacific coast of Mexico, at depths of 546 to 695 fathoms.

It differs from that species in the following particulars :—

The carapace is more granular, and its lobules are capped by blunt conical spinules, *not* smooth tubercles ; and its posterior border is finely and irregularly beaded, *not* dentate.

The front is deeply cut into 3 spines or prongs of almost equal size, *not* into 3 teeth of which the middle one is larger than the others.

The eyes, though very pale, are distinctly pigmented, *not* devoid of pigment.

The inner angle of the wrist of the smaller cheliped is very strongly spiniform, *not* unarmed.

As Mr. Faxon says, *Trachycarcinus* is very closely related to *Trichopeltarium* ; in fact, the relation is so close as to make the separation of the two forms almost doubtful.

CYCLOMETOPA.

Family *Xanthidæ.*

Sub-family *Xanthinæ.*

ORPHNOXANTHUS, Alcock.

Orphnoxanthus, Alcock, J. A. S. B. Vol. LXVII. pt. 2, 1898, p. 127.

Carapace, owing to the inflation of the branchial regions almost quadrilateral in outline and almost concave from side to side, but very decidedly convex fore and aft, broad, the regions well defined but not to any great extent areolated.

Fronto-orbital border a little more than half the greatest breadth of the carapace in extent. Front about a third the greatest width of the carapace, lamellar, projecting horizontally beyond the orbits, broadly and faintly bilobed. Orbital margin entire: orbits and eyes small.

Antero-lateral border cut into four teeth; postero-lateral borders convergent only in the posterior half; posterior border long.

The antennules fold almost transversely. The basal antenna-joint is very short and only just touches the turned down edge of the front; the flagellum which is very long (between 2 and 3 times the length of the orbit) is lodged in the narrow orbital hiatus.

Owing to the bulge of the outer wall of the efferent branchial canal and the consequent puffing out of the pterygostomian regions, the front edge of the merus of the external maxillipeds is quite transverse or even slightly oblique from without inwards and backwards.

The chelipeds are massive and unequal; the fingers are compressed and pointed. The legs are very slender.

The abdomen of the male consists of 5 segments, the 3rd–5th somites being fused.

Owing to the inflation of the pterygostomian regions the efferent branchial channels are permanently open, but the low crests that define them are confined to the posterior part of the endostome.

This genus appears to represent one of the links between *Galene* and *Xantho.* The single known species comes from the Bay of Bengal, 105–350 fms.

Orphnoxanthus microps, Alcock and Anderson.

Xanthodes microps, Alcock and Anderson, J. A. S. B. Vol. LXIII. pt. 2, 1894, p. 183.

Orphnoxanthus microps, Alcock, J. A. S. B. Vol. LXVII. pt. 2, 1898, p. 128, and Ill. Zool. Investigator, Crust. pl. xxxvi. fig. 8.

Carapace about $\frac{2}{3}$ as long as broad, almost quadrilateral in outline, strongly convex fore and aft, but, owing to the inflation of the branchial regions, a little *concave* from side to side; it is rather closely covered with a very fine short fur, beneath which the surface may be granular or nearly smooth, but the margins are always granular. The regions are all well defined and are slightly tumid: the gastric region is divided into 3 gently tumid subregions, the branchio-hepatic regions are subdivided transversely into three areas, and the fronto-orbital margin is also marked off.

The antero-lateral border is thin and sharp and is cut into four sharp finely granular teeth, the first of which runs by a long nearly transverse margin into the (undefined) angle of the orbit. The front is laminar and projects beyond the supra-orbital margin; it is square-cut and is slightly notched in the middle line, so as to form two broad shallow lobes. The eyes are small and are to a variable extent deficient in pigment.

The chelipeds are unequal—very much more so in the male than in the female: the arm to a variable extent, the entire surface of the wrist, and the upper border of the hand are scabrous and more or less hairy; the other surfaces of the hand may be smooth and polished, or the outer surface may be to a variable extent granular: the fingers are large, compressed and pointed.

In the male the larger cheliped is about $2\frac{1}{2}$ times the length of the carapace (the hand and fingers forming slightly more than half the length) and nearly half the arm projects beyond the carapace in repose.

The legs are long slender and finely and sparsely hairy: the upper edge of the meropodites is scabrous or closely spinulate.

Colours in spirit; chestnut brown, with blackish fingers. Length of carapace (average) 11 millim., breadth 15 to 16 millim.

In the Indian Museum are 29 specimens from the Bay of Bengal, 105–350 fms.

PLATYPILUMNUS, Wood-Mason.

Platypilumnus, Wood-Mason MS., Alcock, Ann. Mag. Nat. Hist. May, 1894, p. 401: Alcock, J. A. S. B. Vol. LXVII. pt. 2, 1898, p. 232.

Carapace hexagonal—the prominent bilaminar horizontally-projecting front forming the shortest side of the hexagon—thin, depressed, perfectly flat, with the regions and subregions very faintly impressed: the antero-lateral borders are spinate, the postero-lateral are slightly convergent, and the posterior border is long.

Front about a third the greatest breadth of the carapace. Upper margin of orbit spinate, the inner angle of the lower margin acutely spiniform.

The antennules fold transversely. The basal antenna-joint, though of fair length, does not reach the front; the next joint lies loosely in the wide orbital hiatus; the antennary flagellum is long; about twice the major diameter of the orbit.

Buccal cavern quadrangular, very well defined anteriorly; the external maxillipeds do not nearly cover it, but leave the efferent branchial channels permanently widely open; the endostomial ridges that define these last are well defined posteriorly, but do not reach the anterior border of the buccal cavern.

Chelipeds in the female, markedly unequal, fingers long, pointed.

Legs long, slender, compressed, spiny.

As there is only a single female in the Indian Museum, I cannot be sure of the place of this genus in the system. It *probably* belongs to the *Cancroidea*, and should be placed near *Galene*.

Platypilumnus gracilipes, Wood-Mason.

Platypilumnus gracilipes, Wood-Mason MS., Alcock, Ann. Mag. Nat. Hist. May, 1894, p. 401 : Ill. Zool. Investigator, Crust. pl. xiv. fig. 6 : J. A. S. B., Vol. LXVII. pt. 2, 1898, p. 233.

Carapace much depressed, perfectly flat above, with the surface nearly smooth centrally and very finely and closely granular laterally, and with the regions indistinctly defined. The front has the form of a horizontally projecting bilobed lamella, with the free edge sharply and very evenly spinate and the sides turned abruptly downwards. The margins of the orbit are spinulate, the upper margin the more distinctly so, and the lower margin terminates internally in a strong oblique spine, the point of which inclines towards the sharply vertical tooth formed by the already mentioned downfolding of the lateral edge of the frontal lamella.

The antero-lateral borders of the carapace, which are arcuate and are shorter than the postero-lateral, are armed with three large spines, in front of, between, and behind which are several spinules.

The pterygostomian regions are large and inflated, and the branchial apertures, especially the efferent aperture, are large and patulous.

The eyestalks are large and are of moderate length; the corneal region is rather small.

The antennules are long and are transversely folded, their basal joint is large and inflated.

The antennæ are long, their basal joint is slender and free; the second joint lies loosely in the internal orbital hiatus.

The inner edge of the meropodite of the external maxillipeds is convex, with a pair of little spines at the summit of the convexity; the succeeding joint arises at the antero-internal angle.

The thoracic legs are furnished with many spines and long hairs. The chelipeds, which are robust, are unequal; their prismatic arm has all its borders spiny; the short inflated wrist is sharply granular and spinulate in the distal half of its dorsal surface and along the outer edge, while the inner edge bears a pair of rather large spines; the hand is spinulate everywhere in the smaller cheliped, but only in the proximal third of its outer surface in the larger; the fingers also of the smaller cheliped are spinulate on the outer surface, while those of the larger cheliped are smooth; the cutting-edges of the fingers are finely and unevenly toothed.

The other thoracic legs are long, compressed, and slender, and have the meropodite spiny along both edges, the carpopodite and propodite spiny along the front edge, and the dactylopodite styliform.

Colour in the fresh state yellowish red.

Andaman Sea, 188–220 fms. A single female.

Sub-family *Pilumninæ.*

NECTOPANOPE, Wood-Mason.

Nectopanope, Wood-Mason. Ann. Mag. Nat. Hist. March, 1891, p. 261: Alcock, J. A. S. B. Vol. LXVII. pt. 2, 1898, p. 212.

Carapace broad, approaching the quadrilateral, convex fore and aft, the branchial regions so inflated and convex dorsally as to make the transverse plane of the carapace strongly concave in the middle line, the other regions obscurely defined, the surface smooth.

The antero-lateral borders are very much shorter than the postero-lateral, are very thin and sharp, and are cut into teeth of which the first is confluent with the outer orbital angle.

Front broad, a third the greatest breadth of the carapace, straight, square cut, slightly projecting beyond the supra-orbital angle, from which it is sharply cut off by an angular notch, on either side.

Orbits large, with a thin, sharp, prominent margin; a notch internal to the middle of the upper margin, the notch breaking this margin into two curves, one corresponding to the eye-stalk the other to the cornea: eyes large, reniform, on moderately stout stalks.

Antennules folding transversly. The basal antenna-joint is very short, but almost touches the turned down side-edge of the front: the flagellum, which is considerably longer than the major diameter of the larger orbit, springs from the rather broad orbital hiatus.

The buccal cavern is broader anteriorly than posteriorly, and the mouth parts do not nearly reach its front edge, so that a wide and permanent gap is left: the crests of the endostome are not very strong, but the free edge of the endostome corresponding to the efferent branchial channel, on either side, is deeply excavated. The outer wall of the efferent branchial canal forms a strong angular bulge in the pterygostomian region.

The chelipeds in the female are equal; the fingers are compressed and pointed, not hollowed.

The legs are long and slender, the propodite and dactylus of the last pair strongly compressed and a little broadened.

This form is most nearly related to *Eurycarcinus*.

Nectopanope rhodobaphes, Wood-Mason.

Nectopanope rhodobaphes, Wood-Mason, Ann. Mag. Nat. Hist. March, 1891, p. 261; Alcock, J. A. S. B. Vol. LXVII. pt. 2, 1898, p. 213, and Ill. Zool. Investigator, Crust. pl. xxxviii. fig. 6.

Carapace about $\frac{3}{4}$ as long as broad. Front extremely obscurely grooved in the middle line. Antero-lateral border cut into three thin sharp-edged teeth, of which the first is broad and rounded and confluent with the orbit, the second is broad and anteriorly acuminate, and the third almost spiniform.

Chelipeds smooth: in the female they are equal and are a little over $1\frac{3}{4}$ times the length of the carapace: arm with an acute spine near the far end of the upper border; inner angle of wrist acute, spiniform; fingers thin, compressed, pointed and hooked at tip, armed with thin, laciniate teeth, the thumb very broad.

Legs thin, the first three pairs not much shorter than the chelipeds, with long compressed-styliform dactylus: the last pair a good deal shorter, with thin blade-like propodite and dactylus closely fringed with hair.

Colours in spirit uniform yellowish white: in life pink, with a dotted, **V**-shaped, white mark between the gastric and branchial regions.

In the Indian Museum is a single female specimen from off the Godávari coast 98-102 fms.

Nectopanope longipes, which was provisionally referred to this genus by Wood-Mason, who had insufficient material for examination, turns out, now that numerous good specimens have been dredged by the "Investigator," to be a Catometope belonging to the genus *Carcinoplax*.

Sub-family *Eriphiinæ*.

Sphenomerides, Wood-Mason and M. J. Rathbun.

[Sphenomerus, Wood-Mason.]

Sphenomerus, Wood-Mason, Ann. Mag. Nat. Hist. March, 1891, p. 233: Alcock, J. A. S. B. Vol. LXVII., pt. 2, 1898, p. 227: *Sphenomerides*, M. J. Rathbun, Proc. Biol. Soc. Washington, XI. 1898, p. 164.

Carapace transversely oval or subcircular, the front and antero-lateral margins forming together a semicircle; markedly convex in both directions, perfectly smooth, without trace of regions.

Antero-lateral borders shorter than the postero-lateral—a spinule at their point of junction.

Front somewhat deflexed, broad and broadly bilobed. Orbits affording little or no concealment to the eyes, without fissures or sutures: there is a gap between the front and the inner angle of the orbit, in which the antennary flagellum is lodged. The fronto-orbital border, in the adult, is not quite $\frac{4}{5}$ the greatest breadth of the carapace.

The antennules fold nearly transversely: the basal antenna-joint does not reach the front, the flagellum is a good deal longer than the major diameter of the orbit.

The buccal cavern is a little narrowed anteriorly. The crests of the endostome are very faint, but to make up for this the anterior edge of the buccal cavern is puffed out and is very deeply excised on either side of the middle line; the anterior margin of the foliaceous process of the first maxillipeds is also excised to correspond, and so a permanent expiratory orifice is formed, which is very large and prominent beyond the almost transverse anterior edge of the merus of the external maxillipeds.

The chelipeds are stout, very long and not very unequal; the whole of the arm projects beyond the edge of the carapace: the fingers are somewhat compressed and are pointed.

The legs are rather slender.

The abdomen of the male consists of five pieces, the 3–5th somites being rigidly united but without obliteration of sutures.

Sphenomerides trapezioides, Wood-Mason.

Sphenomerus trapezioides, Wood-Mason, Ann. Mag. Nat. Hist. March, 1891, p. 263: Ill. Zool. Investigator, Crust. pl. v. fig. 2 (where the carapace is drawn a little too broad): Alcock, J. A. S. B. Vol. LXVII. pt. 2, 1898, p. 228.

Carapace about $\frac{4}{5}$ as long as broad, convex in all directions, smooth, polished.

The front is about $\frac{3}{5}$ the greatest breadth of the carapace, is obliquely deflexed, and is divided into two rather shallow broadly-rounded lobes the free edge of which is entire.

The supra-orbital angle is not defined, but the dentiform or spiniform angle of the lower edge of the orbit can be seen from above.

The antero-lateral margins form with the front a semicircular curve, each carries three sharp spinules, namely, one at the outer angle of the orbit, one at the junction with the postero-lateral border and one exactly intermediate between the other two.

The chelipeds are a little, but not very remarkably, unequal: the larger one is about $2\frac{1}{2}$ times the length of the carapace. Their surface is smooth and polished. The arm, the whole of which is visible beyond the carapace, has much the same shape as in *Trapezia*, but its anterior border, though serrated, is not expanded; the lower border of the hand is sharp and somewhat dilated posteriorly, as in *Trapezia*: the inner angle of the wrist is rounded, but sometimes carries a small spinule.

The legs are slender smooth and polished, and have a few hairs distally.

Colours in spirit yellowish white, fingers sometimes blackish in their basal half.

Length of carapace of largest specimen 9 millim., breadth 11 millim.

In the Indian Museum are 11 specimens from the Andaman Sea at depths between 130 and 290 fms.

As the name *Sphenomerus* has been in use since 1860 for a genus of Coleoptera, Miss Mary J. Rathbun (Proc. Biol. Soc. Washington, XI. 1898, p. 164) has proposed to alter the name of this genus to *Sphenomerides*.

Family *Portunidæ.*

Goniosoma, A. Milne Edwards.

Goniosoma, A. Milne Edwards, Ann. Sci. Nat. Zool., (4) XIV. 1860, p. 263, and Archiv. du Mus. X. 1861, p. 367: Miers, Challenger Brachyura, p. 180.

Goniosoma hoplites, Wood-Mason.

Goniosoma hoplites, Wood-Mason, Ann. Mag. Nat. Hist. (4) XIX. 1877, p. 422: Alcock and Anderson, J. A. S. B. Vol. LXIII. pt. 2, 1894, p. 184: Ill. Zool. Investigator, Crust. pl. xxiii. fig. 6.

Surface of body and appendages covered with a dense coat of very short fine adherent hairs. Gastric cardiac and branchial regions well defined, the first two tumid, the last inflated, the summit of their convexities with a few clustered granules.

The gastric region is divided into three subregions and is crossed transversely, near the middle, by an almost straight beaded line: each branchial region is crossed transversely by an anteriorly-convex beaded line, which stops at the gastric region: these are the only ridges on the carapace.

Antero-lateral borders cut into six teeth (including the orbital angle) the first five of which are broad and serrated, while the sixth is an acutely salient spine much longer than the others.

The space between the eyes, which is a third the greatest breadth of the carapace (lateral spines not included) is cut into eight bluntish lobes, the middle two of which are the broadest: the two outermost on either side (which belong rather to the orbit than to the front) are the most acute, and are separated from the middle four, that form the front proper, by a deep notch.

There is a single fissure near the middle of the upper margin of the orbit, the second fissure of this margin being a closed suture. The inner angle of the lower edge of the orbit is not dentiform. The epistome is very slightly sunk.

Chelipeds slightly more than twice the length of the carapace, rather more than half their extent being hand, and having the usual squamiform markings. There are 2 or 3 spines on the distal half of the anterior (inner) border of the arm; a very strong spine at the inner angle of the wrist and three small spines on the outer angle; four spines on the hand, one being on the outer surface at the extreme base, the other three grouped in a triangle at and near the distal end of the upper surface.

Colours in life, salmon pink, clouded on carapace.

Length of carapace 24 millim., breadth including spines 45 millim., measured to base of spines 38 millim.

Coromandel coast, 80 to 110 fms.

Sub-family *Carcininæ.*

Benthochascon, Alcock and Anderson.

Carapace subquadrilateral, its breadth being very little greater than its length, depressed, the regions faintly indicated by slight inequalities of level.

Front about a fourth the greatest breadth of the carapace, sufficiently prominent beyond the orbits, cut into three elegant lobes of nearly equal size, the middle one being notched at tip. Antero-lateral borders hardly arched, much shorter than the postero-lateral borders which they meet at a widely obtuse and hardly noticeable angle, cut into four teeth (including the outer angle of the orbit), the hindmost of which is spiniform. Postero-lateral borders slightly convergent. Posterior border rather longer than the fronto-orbital.

Eyestalks short and thick; eyes very large. The orbits are deep; in the upper margin are traces of two obsolescent sutures, and in the lower margin, just below the outer angle, is a shallow and inconspicuous notch: the inner angle of the lower margin is acutely produced.

The antennules fold almost transversely, their basal joint is very large, their fossæ are widely open to their respective orbits.

The antennæ lie loosely in the orbital hiatus: the basal joint is short slender and moveable, and has its antero-external angle produced to form a blunt spinule; the next joint just reaches the turned down edge of the front, the flagellum is considerably longer than the orbit.

The epistome is wide fore and aft, and is well demarcated from the buccal cavern.

Though the expiratory channels are very well-defined grooves, there are no distinct crests on the endostome.

The external maxillipeds do not close the buccal cavern in front, but fall much short of its anterior margin, leaving the expiratory canals permanently open.

Chelipeds massive, somewhat unequal in the female (male unknown): a strong spine at the inner angle of the wrist.

Legs long, stout and compressed, especially as to the merus joints: the dactyli of the first three pairs are styliform: the dactylus and propus of the last pair are broadly foliaceous for swimming.

This is more closely related to *Bathynectes* than to any other genus, but in *Benthochascon* the front is only three lobed, the antero-lateral borders are cut into four teeth only (including the orbital angle), the slender basal antennal joint is shorter, the meropodites of the first three pairs of legs are stouter and more compressed, and the propus and dactylus of the last pair of legs are more foliaceous; also the carapace is longer, much flatter, and owing to the shortness of the antero-lateral borders, different in shape.

Benthochascon Hemingi, Alcock and Anderson. Plate III. fig. 2.

Carapace about seven-eighths as long as broad, its surface very finely granular, its regions faintly marked as inconspicuous swellings separated by shallow and inconspicuous depressions.

The front forms a thin laminar three-lobed projection. The antero-lateral borders are thin, and are not much more than half the length of the postero-lateral: they are cut into four procurved teeth, of which the foremost is the outer orbital angle and the largest, and the hindmost is spine-like and the longest. The posterior border, between the bases of the last pair of legs, is concave.

The eyes are large and full, their major diameter being from an eighth to a sixth the breadth of the carapace: the inner suborbital angle forms a thin tooth that is almost as prominent as the outer lobes of the front.

The chelipeds in the female are about two-thirds as long again as the carapace, and their surface is everywhere quite smooth to the naked eye, the hand—of which about half is formed by the fingers—forms rather more than half their entire length: the inner angle of the wrist is a large acute spine, and there is a spinule on the inner upper edge of the palm just behind the finger-joint: the fingers are curved, compressed, hooked at tip, and sharply toothed.

The first three pairs of legs are nearly twice the length of the carapace, the third pair being slightly the longest: they are smooth and bare. At the far end of the upper border of the merus of all four pairs is a notch and spiniform tooth.

The last pair of legs is not much longer than the chelipeds, and has the carpus shortened and broadened and the next two joints paddle-like—the edges of all these three joints being elegantly plumed.

Two females from the Andaman Sea, one from 405, the other from 185 fathoms.

The larger specimen, which is laden with tiny eggs, has a carapace 48 millim. long and 51 millim. broad.

Colours in life, recorded by Dr. A. R. Anderson: carapace very pale orange above, the anterior edge and first two teeth of lateral margin white; merus joints pale orange above, the colour extending onto the carpus, the rest of the legs chalky white; eyes bluish-grey: eggs brick red.

CATOMETOPA.

Family *Ocypodidæ.*

Subfamily *Carcinoplacinæ.*

CARCINOPLAX, Edw.

Curtonotus, DeHaan, Faun. Japon. Crust. p. 20.
Carcinoplax, H. Milne Edwards, Ann. Sci. Nat. Zool. (3) XVIII. 1852, p. 164.

Carapace subquadrilateral, transverse, convex fore and aft and strongly declivous anteriorly, nearly flat from side to side, often tomentose, the regions not very well defined.

Front broadish—a third to a fourth the greatest breadth of the carapace in extent—straight and square-cut, generally projecting beyond the orbits.

Antero-lateral borders much shorter than the postero-lateral, commonly armed with 2 or 3 teeth (including the orbital angle): postero-lateral borders subparallel.

Orbits of good size, usually with a low entire margin. The antennules fold quite transversely, beneath the front. The antennal flagella are long, sometimes half the length of the carapace, or more: the basal antenna-joint, which lies in the orbital hiatus, does not nearly reach to the front.

The palate is sharply marked off from the transverse rather narrow epistome, and the efferent branchial channels are well defined. The external maxillipeds completely close the buccal cavern: the merus is broad, with its antero-external angle a little expanded, and gives attachment to the palp at its antero-internal angle.

The chelipeds are equal in length, but one hand may be a little stouter than the other: the wrists are broad and have their inner angle strongly pronounced.

The legs are longish, compressed, and are either tomentose or have the terminal joints plumose.

The abdomen in both sexes consists of seven distinct segments: the second and third segments in the male have the lateral angles produced or lobate and occupy the whole width of the sternum.

Carcinoplax longipes (Wood-Mason).

Nectopanope longipes, Wood-Mason, Ann. Mag. Nat. Hist. March 1891, p. 262: Illustrations of the Zoology of the R. I. M. S. Investigator, Crustacea, pl. xiv. fig. 7.

Carapace subquadrilateral, strongly declivous in front of the level of the lateral-epibranchial angles, smooth, slightly tomentose, the regions very faintly marked.

The front, which is not quite a third the greatest breadth of the carapace, is square-cut, prominent beyond the orbits and folded antennules, entire, sublaminar, and obliquely deflexed. The antero-lateral borders, which are not two-thirds the length of the postero-lateral, are oblique and are armed with two procurved spine-like teeth—one hepatic, the other at the lateral epibranchial angle—and with a third very inconspicuous blunt denticle just behind the orbital angle; the postero-lateral borders are subparallel.

The orbits are rather shallow and afford little concealment to the eyes; their edge is entire. The eyestalks are short, thickish, and normally mobile: the eyes are smallish, but are perfectly formed.

The antennules are large and fold quite transversely; their fossæ are open to their respective orbits.

The basal antenna-joint is slender and very short, the next joint reaches the front, the flagellum is slender and is not much less than half the length of the carapace.

The chelipeds are twice the length of the carapace and are quite smooth: the arm has a denticle beyond the middle of the upper border, the inner angle of the wrist has a very strong spine with sometimes an accessory spinule at its base, the hands are somewhat unequal in size.

The legs are long and slender and have the dactylus thickly plumose, and the two preceding joints more scantily hairy: the third pair, which are slightly the longest, are nearly two and a half times, the fourth (last) pair, which are slightly the shortest, are about twice the length of the carapace.

Colours in spirit pinkish white, fingers blackish brown: in life (recorded by Dr. Anderson) pink, with dark brown fingers.

Length of carapace 13 millim., breadth 15 millim.

1 male and 1 small female from the Andamans, 240–220 fms.: 1 small female from the Andamans, 238–290 fms.: 18 specimens of both sexes from the Travancore coast, 430 fms.

This species is a true *Carcinoplax (Curtonotus)*, as I have assured myself by actual comparison with specimens of *C. longimanus* DeHaan, and of another Indian *Carcinoplax* (from the Gulf of Martaban, 53 and 67 fms.) that appears to be only a variety of *C. longimanus.*

From these the present species chiefly differs in its broader and more prominent front and shorter chelipeds.

The removal of this species to *Carcinoplax* does not affect the integrity of the genus *Nectopanope* (typified by *Nectopanope rhodobaphes*), which is a Cancroid genus closely allied to *Heteropanope* and *Eurycarcinus.*

Psopheticus, Wood-Mason.

Wood-Mason, Admin. Rep. Marine Survey of India, 1890-91, p. 20 (name only).

Carapace transverse, perfectly square, moderately convex fore and aft and declivous anteriorly, nearly flat from side to side, the regions hardly indicated.

Front prominent, square-cut, about one-third the greatest breadth of the carapace, its free margin sharp entire and laminar: antero-lateral borders about half the length of the postero-lateral and in the same straight line with them, the junction being marked by a spine.

Orbits shallow, affording no concealment to the eyes; the hollow for the rather slender eyestalk is very distinctly delimited from that in which the large reniform eye rests: the outer orbital angle is dentiform and very prominent.

The antennules fold quite transversely, their fossæ are widely open to the respective orbits.

Basal antenna-joint short and slender, the next joint reaches the front; the flagellum, which lies in the orbital hiatus, is much longer than the orbit.

Epistome sufficiently broad fore and aft, well delimited from the buccal cavern. The ridges of the endostome are faint, but the expiratory canals are well-defined grooves. The pterygostomian regions are acutely carinated by the bulging of the expiratory canals. The external maxillipeds are just like those of *Carcinoplax (Curtonotus)*, but the gap between them and the front of the buccal cavern is somewhat wider.

Chelipeds moderately massive, of normal length, slightly unequal in the male and still less so in the female. Legs long, moderately stout, almost hairless, some of the joints are spiny: dactyli long and slender.

The abdomen of the male consists of seven distinct segments and occupies all the space between the last pair of legs.

This genus is very little different from *Carcinoplax*, *Pilumnoplax*, *Litochira* etc. and perhaps ought rather to be regarded as a subgenus of *Carcinoplax* than as an independent genus. All that separate it from *Carcinoplax* (e. g. *C. longimanus*) are the perfectly square carapace, the less distinct endostomial ridges, and the spiny merus joints and slender hairless dactyli of the legs.

Psopheticus stridulans, Wood-Mason.

Psopheticus stridulans, Wood-Mason, Illustrations of the Zoology of the Investigator, Crustacea, pl. v. fig. 1 (1892): Alcock, Ann. Mag. Nat. Hist., May 1894, p. 402.

Carapace three-fourths as long as broad, smooth and polished, crossed transversely in its posterior half by a broad groove which is continued obliquely across the pterygostomian regions to the angles of the mouth.

A thin sharp prominent tooth at the outer orbital angle, and an obliquely-prominent spine at the junction of the antero-lateral and postero-lateral borders.

The subocular and subhepatic regions are inflated, and together form a granular eminence against which a strong spine on the upper border of the arm can be brought to play, producing a sound. Hence the names *Psopheticus* and *stridulans*.

The major diameter of the reniform eye is between a sixth and a seventh the breadth of the carapace: though the orbit does not conceal the eye its edges are well and cleanly cut.

The chelipeds in the adult male are a little more, in the adult female a little less, than twice the length of the carapace, and are smooth and polished, as also are the legs. The arm has a strong upstanding claw-like tooth near the middle of its upper border, one or two spinules near the far end of the outer border, and a spinule near the far end of the inner border: the wrist has both the inner and the outer angles spiniform.

The third pair of legs, which are slightly the longest of the four, are rather more than two-and-a-half times the length of the carapace. In all, the anterior edge of the meropodites is armed with spines and the same edge of the carpopodites with spinules—these being least numerous and least distinct in the case of the first pair.

Colours in glycerine: chelipeds and legs rather dusky red; carapace dusky red behind the transverse groove—which forms a very sharply-defined red band —livid red, or almost violet, in front of it; eyestalks almost purple, eyes purplish-black. Eggs in life magenta.

The carapace of the largest male is 15 millim. long and 20 millim. broad.

Only known, so far, from the Andaman Sea: 2 males and a female from 173 fms., 2 males and a female (Types of the species and genus) from 188–220 fms., 7 females (3 with eggs) from 185 fms., a male and 4 females from 370–419 fms.

Pilumnoplax, Stimpson.

Pilumnoplax, Stimpson, Proc. Acad. Nat. Sci. Philad. 1858 (1859) p. 93: Miers, Challenger Brachyura, p. 225.

Closely related to *Carcinoplax* from which it appears to differ only in the following particulars:—

(1) The carapace, instead of being convex fore and aft, is flat and depressed.

(2) The front is somewhat broader and the orbits are deeper.

(3) The ridges that define the efferent branchial channels are rather less distinct.

Pilumnoplax Sinclairi, Alcock and Anderson. Plate III. fig. 1.

Carapace subquadrilateral, much depressed, a little more than three-quarters as broad as long, very finely frosted, perfected bare, the regions fairly indicated.

Front horizontal, slightly prominent, square cut, grooved but not distinctly notched in the middle, more than a third the greatest breadth of the carapace; its free edge is turned vertically downwards to form a narrow concave facet with raised margins.

The antero-lateral borders are not much more than half the length of the postero-lateral: they are thin and sharp, and are cut into three teeth, of which the first is broad and somewhat emarginate and the other two are acute. On the postero-lateral borders, just behind the junction with the antero-lateral, is a denticle.

The eyes are small but well-formed, and are freely movable. The orbits conceal the retracted eyes to dorsal view: their upper margin is fissured near the middle, and the lower margin is slightly excavated just below the outer angle: the inner angle of the lower margin is not prominent, though dentiform.

The antennules fold transversely, and their fossæ are freely open to their respective orbits.

The basal antenna-joint is short and slender: the next joint reaches the front: the flagellum, which arises in the orbital hiatus is about twice the length of the orbit.

The outer maxillipeds completely close the buccal cavern.

The chelipeds in the female (male unknown) are unequal, the larger one being not quite twice as long as the carapace: their surface, under the lens, is finely frosted: the inner angle of the wrist is strongly pronounced and is capped by a pair of acute teeth.

Legs moderately stout, unarmed, smooth, almost hairless: the third pair, which are somewhat the longest, are about two-and-a-half-times the length of the carapace. The dactyli are compressed-styliform.

Colours in spirit french-grey, fingers much darker grey.

A single female specimen, from off the Travancore coast 430 fms., has the carapace 13 millim. long and 16 millim. broad.

This species is closely related to *Pilumnoplax heterochir* (Studer) Miers, but is distinguished from it by the entire and more prominent front, by the absence of transverse markings on the carapace, by the longer legs, and by the smoothness of the chelipeds and legs.

From *Pilumnoplax abyssicola* Miers, which it also closely resembles, it is distinguished by the smooth carapace (to the naked eye), by the turned-down milled edge of the front, by the spinule on the postero-lateral border, by the fissured upper-margin of the orbit, and by the double spine at the inner angle of the wrist.

Subfamily *Rhizopinæ.*

CAMATOPSIS.

Carapace deep, rudely sub-semicircular, hardly broader than long, strongly convex fore and aft and declivous anteriorly, nearly flat from side to side across the branchial regions: its antero-lateral borders short sharp and entire, its postero-lateral borders long sharpish and parallel or slightly divergent: its only markings are two longitudinal grooves, hardly visible on the undenuded carapace, that mark off the epibranchial regions.

Front considerably less than a fourth the greatest breadth of the carapace, obscurely bilobed.

Orbits large, deep, the upper margin entire and cut in the anterior border of the carapace: eyestalks large, tumid, conical, almost immovably fixed in the orbits: eyes reduced to a speck of pigment placed on the under surface of the tip of their stalks.

Antennularly fossæ widely open to their respective orbits, small, and filled entirely by the basal antennulary joint, *to the complete exclusion of the large flagellum:* the interantennulary septum, though very narrow, is complete.

The small basal antenna-joint is wedged in between and beneath the eye-stalk and antennule, the second joint hardly reaches to the front, the flagellum is large and considerably longer than the orbit.

The epistome is of considerable width fore and aft, especially at its middle. The buccal cavern is square, though rather broader in front than behind, and is almost entirely covered by the external maxillipeds. These have the merus as long as, and markedly broader than the ischium, owing to the semilunar expansion of the outer border of the merus: the palp, which is of good size, is jointed to the antero-internal angle of the merus.

The chelipeds are moderately massive and have their movements of abduction and extension somewhat restricted, in the male the hands are unequal. The arm is short and trigonal, the wrist rather long and crooked.

Legs sufficiently long and stout, the penultimate pair being the longest; their dactyli are sharply trigonal and elegantly plumose: the last pair are subdorsal and have the dactylus slightly curved and compressed.

The abdomen of the male, which is four-jointed: does not nearly fill the space between the last pair of legs.

Between the 4th and 5th segments of the sternum, in the male, is intercalated a long narrow plate that appears to cover the external genital ducts.

This seems to be more closely related to *Xenophthalmodes* than to any other genus.

Camatopsis rubida, Alcock and Anderson. Plate IV. fig. 3.

Carapace very finely granular when denuded. The narrow front and the antero-lateral borders form a semicircular curve: the postero-lateral borders are slightly divergent, the greatest breadth of the carapace being between the bases of the penultimate pair of legs. The tumid anterior (true inner) borders of the eyestalks bulge beyond the orbital concavities of the anterior border of the carapace.

The efferent branchial canals cause an angular bulging or carination of the pterygostomian regions.

The chelipeds are unequal in the male (female unknown), the longer one being about 1¾ times the length of the carapace. They are unarmed. In the larger hand the fingers meet only at tip and are finely toothed in the distal half only, being rather deeply notched in the basal half, while on the inner surface of the movable finger is a curious truncated spine. In the smaller hand the fingers meet throughout their extent and only the immovable finger is distinctly toothed, one or two of its teeth being enlarged.

The first and last pair of legs are about 1⅔ times, the second and third pair are about twice, the length of the carapace. In the last pair of legs the terminal joints are more strongly ciliated, and the dactylus is slightly curved and compressed as for swimming.

Colours in spirit rich chocolate brown. Animal entirely covered with velvet.

Three males from the Andaman Sea, 194 fathoms. The carapace of the largest is 9 millim. long and 10 millim. broad.

Hephthopelta n. gen.

Carapace very deep, inflated, rudely semicircular, about as long as broad, convex fore and aft and vertically deflexed anteriorly, all its borders entire and all except the posterior tumid, the cardiac and branchial regions well delimited.

Front considerably less than a third the greatest breadth of the carapace, bilobed, vertically deflexed.

Orbits small, shallow, excavated in the vertically-deflexed anterior border of the carapace, not concealing the eyes. Though the eyes are small and their stalks immovably fixed, they are well formed, well defined and well pigmented.

The antennulary fossæ are widely open to their respective orbits, and are filled by the basal antennulary joints, to the exclusion of the flagella: the inter-antennulary septum, though very narrow, is complete.

The basal antenna-joint is small, slender, and does not nearly reach the front; it lies between the eyestalk and antennule: the flagellum, which arises in the orbital hiatus, is hardly longer than the orbit.

The epistome is of considerable width fore and aft. The buccal cavern is square, though slightly narrower in front than behind: the excurrent branchial canals are well defined. The external maxillipeds, which completely cover the buccal cavern, have the merus shorter and slightly narrower than the ischium and somewhat oval in shape, and the palp jointed to the antero-internal angle of the merus and of good size.

The legs are all long and slender and end in a slender dactylus: the last pair are subdorsal.

The chelipeds are lost in the single specimen obtained, which is a female.

Hephthopelta is closely allied to *Camatopsis*, but differs from it (1) in the form of the carapace, which is inflated and subspherical and has the fronto-orbital region vertically deflexed, (2) in the well formed eyes, (3) in the small antennary flagella, (4) in the narrower and differently shaped merus of the external maxillipeds, and (5) in the longer and more slender legs.

Both it and *Camatopsis* connect the *Rhizopinæ* with the *Pinnoteridæ*.

Hephthopelta lugubris, n. sp. Plate IV. fig. 2.

Carapace as long as broad, roughly semicircular or semiglobose, of thin texture, its surface very finely frosted and somewhat pubescent.

The fronto-orbital region is vertically deflexed and almost invisible in a dorsal view.

Epibranchial and cardiac regions tumid, circumscribed by deepish grooves.

Legs subcylindrical, with a finely frosted and pubescent surface: the third pair, which are slightly the longest, are about $2\frac{3}{4}$ times the length of the carapace: the posterior (lower) border of the merus of the first two pairs is spinulose.

Colours in spirit, light yellow, eyes black.

A single female, without chelipeds, from the Andaman Sea, 490 fms. The carapace is 8 millim. long, and the same in breadth.

Family *Ptenoplacidæ*.

Carapace flat, depressed. Front extremely narrow, projecting freely far beyond the interantennulary septum. Orbits imperfect. No distinct antennulary fossæ. Epistome none (linear): buccal cavity broad, subquadrate: external maxillipeds subpediform and affording almost no concealment to the underlying (*i. e.* overlying) parts, the palp articulating at the summit of the narrow merus. In neither sex does the abdomen nearly cover the sternum between the bases of the fourth pair of peræopods.

[Fifth pair of peræopods are subdorsal in position, are reduced in size, and arise near the middle line of the body.] Type *Ptenoplax notopus.*

At first sight, from its general shape, from its elongate third pair of trunk-legs and its almost rudimentary notopodal fifth pair, from its extremely incomplete orbits, from the absence of antennulary fossæ, and from the curiously small and slender external maxillipeds, Homolid affinities are suggested; but that this singular form has nothing to do with the *Homolidæ* is shown: (1) by the position of the openings of the oviducts, which is typically Cancroid; (2) by the form and position of the openings of the efferent ducts of the male, which are typically Catometopan; and (3) by the number and disposition of the branchiæ, of which there are only six on each side.

In the number and arrangement of the branchiæ, as well as in the position and degenerate form of the fifth pair of legs and the obsolete epistome, it might be supposed that it had affinities with the *Dorippidæ* (*Dorippe* and *Ethusa* more especially). That this is not the case is shown (1) by the position, above indicated, of the genital openings of the male; (2) by the great broad buccal orifice, which is only very partially covered by the maxillipeds; (3) by the form of the carapace, which is broad, and completely covers the thorax; (4) by the form of the antennules, which are not obliquely or almost vertically folded in distinct fossæ as they are in *Ethusa* and *Dorippe*; and (5) by the form of the sternal plastron, which in our new form is a broad pentagonal plate as in many Ocypodoids.

It is a true Catometope, but of an archaic type.

PTENOPLAX, Alcock and Anderson.

Archæoplax, Alcock and Anderson, J. A. S. B. Vol. LXIII. pt. 2. 1894. p. 180.
Ptenoplax, Alcock and Anderson, Ill. Zool. Investigator, Crust. pl. xv. 1895.

Carapace transverse, greatly depressed, with the front very narrow, and declivous, yet forming a distinct rostrum (*i. e.*, its front border is not fused with the epistome, but is free). Abdomen in both sexes narrow, not nearly co-extensive in breadth with the sternum between the penultimate pair of trunk-legs.

Orbits and antennulary fossæ very imperfect, hardly more perfect than in *Homola*, and formed on somewhat the same plan as they are in the *Dromidæ*. That is to say, there is a shallow common orbito-antennulary fossa—into which, however, the antennules fold transversely, and entirely below, not beside, the eyes—and the lower boundary of this fossa is formed by the basal joint of the antennule, the basal joint of the antenna, and an external tooth. The eyestalks are moderately long, are slender and tapering, and do not nearly fill their region of the orbit: the eyes are small.

Antennules well developed, transversely folded on the inflated basal joint, *which is free and exposed from its origin.* Antennal peduncles arising external to, and in the same plane with, the antennules: the flagella long.

Buccal opening much wider in front than behind, not nearly covered by the short slender external maxillipeds: efferent branchial channels well defined, anteriorly produced and patulous: epistome linear: the palp of the external maxillipeds articulates with the apex of the narrow meropodite.

Chelipeds unequal in the male, sub-equal in the female: first, second and third pairs of legs long, slender and compressed (the second pair the longest), with long sabre-shaped dactyli. *Last pair of legs reduced to feather-like rudiments, arising close together, high up, almost on the back.*

Genital ducts of the male opening at a distinct tubercle on the base of the fifth pair of legs, the tubercle being embedded in a notch in the posterior border of the sternum.

Six gills on either side.

Ptenoplax notopus, Alcock and Anderson.

Archæoplax notopus, Alcock and Anderson, J. A. S. B. Vol. LXIII. pt. 2, 1894, p. 181, pl. ix. figs. 3, 3*a*-*b*.
Ptenoplax notopus, Alcock and Anderson, Ill. Zool. Investigator, Crust. pl. xv. figs. 2, 2*a*-*b*.

Carapace extremely flat and depressed, transversely oval, with the anterior and antero-lateral margins slightly concave; its surface punctate beneath a shaggy reddish fur.

The front proper is extremely narrow—about one-fourteenth the greatest breadth of the carapace—and is deflexed with the tip free and horizontal, the tip also being slightly expanded and bilobed just as in *Macrophthalmus*.

The orbital borders of the carapace, which together are half the greatest width of the carapace, are concave on either side of the front, each concavity being interrupted near the middle by a small projection: the antero-lateral borders are short and oblique, are broadly concave, and are rather acutely produced at their junction with the anterior margin: the postero-lateral borders, which constitute four-fifths or more of the lateral extent of the carapace, are convex, and form a small tooth at their angular junction with the antero-lateral borders: the posterior border is raised and gently convex. The inflated branchial regions are fairly well delimited from the gastro-cardiac regions.

Two remarkable almost straight sutures cross the carapace from side to side: the anterior at the level of the junction of the antero-lateral with the postero-lateral borders, the posterior at the middle of the cardiac region. These sutures are remarkably distinct, equally from the exterior and from the interior of the carapace.

The side walls of the carapace, though low, meet the dorsal surface almost at a right angle. The pterygostomian regions are deeply grooved or creased transversely in the neighbourhood of the large afferent branchial orifices. The sternum in both sexes is widely pentagonal.

The orbits are remarkably incomplete, their inferior border being formed only by a large acute lamelliform spine and by the basal joint of the antennule and of the antenna.

The eye-stalks are long (their length being contained 6 or 7 times in the greatest breadth of the carapace), slender, tapering, and slightly bent: the eyes are small and hemispherical.

The antennules have the basal joint hugely inflated, globular, quite free and exposed from its origin, and freely mobile: the second and third joints, which are long and slender, fold transversely on the base of the first.

The antennæ arise just below the infra-orbital spine, and outside and in the same line with, the antennules: their flagellum is half the length of the carapace.

The buccal cavity is considerably wider in front than behind: the external maxillipeds are so small and slender as to leave completely exposed the mandibles, the wide endostome, and a part of the wide and produced efferent branchial channels.

The epistome is linear. The fourth joint of the external maxillipeds arises from the apex of the small oval third joint.

All the trunk-legs are thickly fringed with a shaggy reddish hair.

The chelipeds are subequal in the female, but are unsymmetrical in the male: their length, half of which is formed by the hand, slightly exceeds the breadth of the carapace: both hands in the female, and the smaller hand in the male, are elongate compressed and sharp-edged, and have the fingers curved compressed acute, slightly excavated on the inside, and indistinctly dentate along the opposed edges: the larger hand of the male has the palm inflated.

Of the legs, the 2nd pair is the longest, measuring rather more than twice the greatest breadth of the carapace: all are slender compressed and quite smooth, and all end in long sharp sabre-shaped dactyli.

The last pair of trunk-legs are quite unique in form and disposition: they arise quite close to the middle line of the body and high up, almost on the back; they are short, being considerably less than the breadth of the carapace in length, and are very slender and flexible; and they are so thickly fringed with shaggy hairs as to look like feathers.

The abdomen in the male consists of 5 separate pieces—the 3rd–5th segments being coalescent: its breadth opposite the penultimate pair of trunk-legs is about one-third that of the sternum at the same point. In the female the abdomen consists of 7 separate segments, and its breadth opposite the penultimate pair of trunk-legs is half that of the sternum at the same level. The genital openings in the female have the usual position on the sternum: in the male they are placed at the summit of a prominent tubercle situated at the antero-internal angle of the basal joint of the 5th pair of legs, the tubercle being embedded in a notch in the posterior border of the sternum.

The carapace of an average egg-laden female is nearly 17 millim. long, and nearly 22 millim. broad. Males are somewhat smaller.

Colours chestnut brown, carapace lighter: eggs scarlet.

Bay of Bengal, Coromandel coast 100-250 fms.; Andaman Sea, 185 fms.

Family *Pinnoteridæ.*

Pinnoteres, Latreille.

Pinnotheres, Latreille, Hist. Nat. Crust. et Ins. VI. 78 and in Cuvier, Règne An. 2nd ed. 1829, p. 48: Lamarck, (Hist. Nat. An. sans V. (2nd edit.) Vol. V. p. 410): Bosc, Hist. Nat. Crust., I. p. 239: Leach, Malac. Pod. Brit.: Desmarest, Consid. Gen. Crust. p. 116: Milne Edwards, Hist. Nat. Crust. II. 30, and Ann. Sci. Nat., Zool., (3) XVIII. 1852, p. 138 and (3) XX. 1853, p. 216: De Haan, Faun. Japon. Crust. p. 34: Dana, U. S. Expl. Exp. Crust. pt. I. p. 378: Bell, British Stalk-eyed Crust. p. 119. Miers, Challenger Brachyura, p. 275. Ortmann, Zool. Jahrb. Syst. etc. VII. 1894, p. 698: O. Bürger Zool. Jahrbacher, Syst. etc. VIII. 1895, p. 362: Adensamer, Ann. Nat. Hist. Hofmus. Wien., 1897, p. 105.

Pinnoteres abyssicola, Alcock and Anderson. ♀

Carapace as long as broad, circular, smooth: front rather prominent, about one-fifth the greatest breadth of the carapace. The whole of the eyes and eye-stalks and almost the whole of the orbit are visible in a dorsal view. The eyes are well developed but very pale. The dactylus of the external maxillipeds is styliform, and is inserted at the end of the preceding joint. The lower border of the immobile finger is fringed with fine hairs. The legs are slender: the 2nd and 3rd pairs are both about $1\frac{1}{2}$ times as long as the carapace, and have the dactylus slightly longer than it is in the other two pairs.

A single female with eggs, and with a carapace about 8 millim. in diameter, was taken by Dr. A. R. S. Anderson from a living individual of a large species of Lamellibranch (*Lima indica*, E. A. Smith) dredged off the coast of Travancore at a depth of 430 fms.

It is interesting to notice that this species is quite like any other *Pinnoteres* and has apparently undergone no further modification to fit it for deep-sea life.

Mr. E. A. Smith, who has kindly named the host species for me, writes that "it is exceedingly close to the Norwegian *Lima excavata*............*Lima goliath* from Japan is another near relation. It is questionable whether they do not all belong to one widely distributed form."

Mr. Smith, to whose voluntary kindness we are indebted for numerous interesting reports—published in the *Annals and Magazine of Natural History* for 1894 and succeeding years—on the Investigator Deep-sea Mollusca, has already remarked the "close similarity" of certain of them "to species which occur in the North Atlantic."

Since the foregoing was sent to press I have received a copy of Miss M. J. Rathbun's *Brachyura of the Florida Keys* (Bull. Lab. Nat. Hist. Iowa, June, 1898, pp. 250–294, pl. i-ix), in which are to be found some remarkable corroborations of the views expressed in the Introduction to this Report as to the Atlantic affinities of the fauna of the moderate depths of Indian Seas.

For instance, *Thyrolambrus astroides* Rathbun (1894), dredged by the "Albatross" off Havana in 67 and 189 fathoms, is believed by Miss Rathbun, who has had the figure published in the "Investigator" *Illustrations* to guide her, to be identical with *Parthenope (Parthenomerus) efflorescens* Alcock (1895), dredged by the "Investigator" in the Andaman Sea at 36 fathoms. Miss Rathbun also records the species from Mauritius.

Again, *Pilumnoplax americana* Rathbun (June 1898) dredged off the coast of Florida in 70–110, 116, and about 200 fathoms, is identical with the *Pilumnoplax Sinclairi* here described (p. 74), from 430 fathoms off the Travancore coast.

Further, *Chasmocarcinus* Rathbun, from off Trinidad, 31–34 fathoms and off the Bahamas in 97 fathoms, seems to differ but slightly from the *Camatopsis* of page 75 of this Report.

Also, after this paper was sent to press I discovered a new species of *Mursia* (*M. aspera*) that had been stowed away by some well-meaning attendant. The discovery was made just in time to enable me to insert the description of the species in its proper place but not soon enough to allow for the correction of the table of the percentage composition of the Indian Brachyurous fauna on p. 3.

ADDENDA

The three following deep-sea species, which have just come in from the ship, as part of the collection of the present season (1898–99), have to be added.

OXYRHYNCHA.

Family *Maiidæ*: Sub-family *Inachinæ*.

GRYPACHÆUS, Alcock.

Grypachæus, Alcock, Journ. Asiatic Soc. Bengal, Vol. LXIV. Pt. ii. 1895. p. 177.

Carapace triangular, spiny, separated from the frontal region by a post-ocular "neck," not concealing the first two abdominal terga even in the male.

Rostrum spiny: composed of two short divergent spinelets, with a strong median deflexed (interantennulary) spine, not visible from above. Eyes laterally projecting, movable, but not sufficiently retractile to be ever concealed. Small supra-ocular and post-ocular spines are present as part of the general spinature. Antennæ dorsally exposed from the basal joint of the peduncle, which joint is long slender cylindrical and spiny and is not intimately fused with the neighbouring parts. External maxillipeds with the merus elongate, much narrower than the ischium, and not much broader than the carpopodite. Legs hairy and spiniferous. Abdomen six-jointed in ♀, curiously short and truncated in ♂.

Branchial formula apparently as is *Encephaloides*.

Grypachæus hyalinus (Alcock & Anderson.)

Achæus hyalinus, Alcock & Anderson, J. A. S. B., Pt. ii. 1894. p. 205.
Grypachæus hyalinus, Alcock, J. A. S. B., Vol. LXIV. Pt. ii. 1895. p. 177. pl. iii. figs. 4. 4a.

Carapace sub-triangular, thin, vitreous, spiny especially in its anterior half: the regions well delimited, and the post-ocular portion constricted to form a "neck." The gastric and branchial regions, but more especially the latter, are particularly convex.

The rostrum, as seen from above, ends in two short spines, each of which has a spine or two at its base; but from in front or from below it shows as a strong vertically deflexed (interantennulary) spine.

The eyes are large; and the long eye-stalks, which bear one or two tubercles on their front surface, are movable backwards, and are exposed from their base in all positions. The antennæ are visible, dorsally, from the end of the basal joint of the peduncle, which joint is long, slender, cylindrical and spiny.

The external maxillipeds are large, hairy, and almost pediform, owing to the narrowness of the merus and the coarseness of the palp.

The chelipeds in the male are half again as long as the carapace and rostrum and are very much more massive than the legs, and in the female are as long as the carapace and rostrum and not much more massive than the legs: all the joints except the fingers are spiny. The fingers are much shorter than the palm, which in the male is considerably inflated.

The first three pair of legs are spiny and hairy, the hairs of the posterior edge being remarkably long, stiff, close-set, and regular: the first pair are the longest, being in the male rather more, in the female rather less, than twice the length of the carapace and rostrum: the next two pair decrease in length successively. The last pair, which are also hairy, are the shortest of all, are subdorsal in position, and are subchelate, the propodite having its near end dilated to receive the folded back dactylus:

the apposed edge of the dactylus being minutely, that of the propodite being sharply and conspicuously, spinate. Except that the dactylus is nearly as long as the propodite, the last pair of legs is very like that of *Homola*.

The largest male has a carapace 18 millim. in extreme length and 11 millim. in greatest breadth.

Off the Malabar coast, 68–148 fathoms.

'Two females of this species were dredged off Trincomalee in 28 fathoms some years ago; but the characteristic expansion of the branchial chambers shows that the species belongs to the deep-sea fauna.'

Subfamily *Pisinæ*.

SPHENOCARCINUS, A. Milne Edwards.

Sphenocarcinus auroræ, n. sp.

Closely resembles *S. cuneus* (see p. 50), but differs in the following particulars:—

The rostrum, measured from the anterior extremity of the post-ocular process, is in the male about three-fourths, in the female from three-fifths to three-fourths, the length of the rest of the carapace: in both sexes it is equally distinctly bifid at the end.

The pre-ocular processes end subacutely, and are therefore much more spine-like than in *S. cuneus*.

The tubercles of the carapace, though they have the same position and much the same shape, are smaller, and are therefore separated by much wider channels: the gastric tubercle instead of touching the cardiac tubercle, is widely separated from it, and the cardiac tubercle is heart-shaped rather than triangular.

As regards the male sex only: the pair of carinæ on the dorsal surface of the carpopodites of the chelipeds and legs are blunt, the carina on the dorsal surface of the meropodites is almost indistinguishable, and the arms of the chelipeds are very much stouter than the corresponding joints of the gressorial legs.

Colours of fresh spirit specimens: light saffron-pink suffused with rose-pink, the rostrum almost crimson.

Twenty-seven specimens from off the Travancore coast, 224–284 fms.

Length of carapace and rostrum 19 millim. ♂ and ♀.

,, ,, rostrum alone 8 ,, ,,

Greatest breadth of carapace 13·5 ,, ,,

In some females the rostrum is relatively shorter.

CYCLOMETOPA.

Family *Xanthidæ*. Subfamily *Galeninæ*.

Geryon, Kröyer.

Geryon, Kröyer, Nat. Hist. Tidskr. (1) I. 1837, p. 20: ?Miers, Challenger Brachyura p. 223; de Man, Notes Leyden Mus. XII. 1890, p 69: A. Milne Edwards et Bouvier, Résult. Camp. Sci Hirondelle, Brachyures et Anomoures, p. 41 (Monaco, 1894).

Chalæpus, Gerstaecker, Archiv. f. Naturges. XXII 1856, i p 118.

Carapace subquadrilateral with the fore-part rounded, very little broader than long, convex (sometimes strongly so) fore and aft, slightly convex from side to side across the branchial regions, the surface more or less pitted or finely eroded, the regions obscurely defined.

Antero-lateral borders rather shorter than the postero-lateral, moderately arched, with from 3 to 5 teeth or acuminate lobes (including the outer orbital angles).

Fronto-orbital border about half the greatest breadth of the carapace. Front horizontal, or obliquely deflexed, about one-fifth the extreme breadth of the carapace, quadrilobed or quadridentate, the two middle teeth slightly the more prominent.

Supra-orbital margin with two suture lines or fissures, of which the outer one is sometimes hardly discernible: inner angle of lower border of orbit usually dentiform and prominent.

The antennules fold nearly transversely. The antennæ occupy the widely-open orbital hiatus: The basal joint is short and is not fixed down, the second joint reaches the front, the flagellum is long.

Buccal cavern square, the efferent branchial channels well defined by ridges: external maxillipeds large, completely closing the buccal cavern, the anterior border of the merus somewhat oblique and arched.

Chelipeds subequal, massive, fingers strong and pointed.

Legs long, stoutish, more or less compressed, ending in stout, bare, styliform dactyli.

Abdomen in both sexes seven-jointed and completely covering the sternum at its base.

Geryon affinis, Milne Edwards and Bouvier.

Geryon affinis, A. Milne Edwards et E. Bouvier, Résult. Camp. Sci. Hirondelle, Brachyures et Anomoures, p. 41, pl. i. (Monaco, 1894).

One specimen, a female with a carapace nearly 4·5 inches long and 5 inches broad, from off the Travancore coast, 224–284 fms. It is absolutely identical with the species dredged off the Azores at about 310–770 fathoms, and unmistakeably described and figured by MM. Milne Edwards and Bouvier. To those who know how many species are common to the like depths of the two regions its occurrence here will cause no surprise.

Carapace slightly broader than long, little convex, its surface rugulose or pitted, especially on the branchial regions, its regions fairly well defined, the gastric region convex and obscurely divided into three sub-regions.

Front about one-fifth the greatest breadth of the carapace, obliquely deflexed, lamellar, cut into 4 teeth of which the middle two are the most prominent.

Antero-lateral borders with five prominences or teeth (including the outer orbital angle) of which the 1st, 3rd, and 5th are the most pronounced: a broad low ridge extends from the last tooth right round the inner limit of the branchial region on either side.

Orbits large, not completely filled by the eyes: of the two breaks in the upper margin the inner is a distinct fissure while the outer is a mere crease: the inner angle of the lower border is prominent and acutely dentiform.

The external maxillipeds fall a little short of the edge of the epistome, the surface of their ischiopodite, like that of the neighbouring part of the pterygostomian regions is rugulose or pitted.

Chelipeds slightly unequal: all surfaces of the hand, the outer surface of the wrist and part of the outer surface of the arm are marked with low sub-squamiform rugulosities: the upper border of the arm ends, some distance behind the end of the joint, in a spine, and the inner angle of the wrist is strongly spiniform: the fingers are longer than the palm and their apposed teeth are strong and interlock closely.

The gressorial legs are stout, and their surfaces are pitted: they end in stout styliform dactyli which are grooved along both the dorsal and the ventral surfaces.

The colours in life are admirably shown in the figure given by MM. Milne Edwards and Bouvier.

www.ingramcontent.com/pod-product-compliance
Lightning Source LLC
Chambersburg PA
CBHW020257090426
42735CB00009B/1116
9783337161163